the joy of keeping a
root cellar

the joy of keeping a
root cellar

CANNING, FREEZING, DRYING, SMOKING, AND PRESERVING THE HARVEST

Jennifer Megyesi

photography by Geoff Hansen

Skyhorse Publishing

Skyhorse Publishing books may be purchased in bulk at special discounts for
sales promotion, corporate gifts, fund-raising, or educational purposes. Special
editions can also be created to specifications. For details, contact the Special
Sales Department, Skyhorse Publishing, 555 Eighth Avenue, Suite 903,
New York, NY 10018 or info@skyhorsepublishing.com.

www.skyhorsepublishing.com

10 9 8 7 6 5 4 3 2 1

Library of Congress Cataloging-in-Publication Data

Megyesi, Jennifer Lynn, 1963—The joy of keeping a root cellar : canning, freezing, drying, smoking, and preserving the harvest /
Jennifer Megyesi ; photography by Geoff Hansen.
    p. cm.
 ISBN 978-1-60239-975-4 (pbk. : alk. paper)
1. Vegetables—Storage. 2. Fruit—Storage. 3. Root cellars.
4. Food--Storage. I. Title.
TX612.V4M38 2010

641.4'8—dc22
                              2010010322

Printed in China

For **Grandpa**, for his stories of the railroad and for walking with my sister and me to get Cokes in real glass bottles for 30 cents at the corner drugstore. For Grandma, who let me crawl into bed with her when I was a child and comforted me back to sleep. For **Grandpapa**, who had no patience with my questions and only wanted me to figure out for myself the importance of self-sustenance. And to **Grandmama**, who gave me the courage to follow my dreams. While only memories of them remain, tucked away in my past as a child, my grandparents taught me that traditions are important enough to be carried out and kept alive in the present.

CONTENTS

Myla Rogers, 9, a fourth-grader at Randolph Elementary School, carries winter squash she found in a harvested field at the author's farm Fat Rooster Farm in Royalton, Vermont. The students in Linda Garrett's third- and fourth-grade class are part of a school-wide project for which they are studying a particular farm through the year.

Preface

THE ART OF FOOD PRESERVATION was as important as spoken and written language for bringing about the means of modern civilization. Enzymes and microorganisms that naturally occur within food as well as the surrounding atmosphere begin to change its composition immediately after it has been slaughtered or harvested. Preservation techniques like drying, fermenting, freezing, and canning preserve food by preventing these processes from occurring. With the advent of food preservation, no longer was it necessary to consume the food directly after it had been killed or harvested; instead, it gave the nomad a surplus to rely upon for later use. Able to make roots in one place, human beings were free to form social structures and bonds—villages, cities, and communities. It became possible for inhabitants to interact and share living experiences with one another on a daily basis throughout the seasons.

More close to home are my childhood memories of the basement in my Grandmama's house outside Detroit, Michigan. The home, in a sea of homes laid out in snaking patterns of uniform shapes and colors (on streets that crisscrossed so much that I once got lost on my bicycle trying to catch the ice-cream truck while it peddled its summer treats), was textbook suburbia, and about as far away from a nomadic living style as you could go. But in that basement, in one tucked-away corner, there were shelves lined with jars of preserved foods.

Memories of a being a ten-year-old can quickly become a muddled stream of fantasy and fact, but some of the most vivid images that remain with me now are of the smell of drying tobacco leaves, sour cherries, pigs' feet preserved in gelatin, jams, jellies, rising pastry dough, and fermenting wines. It was cool down there, cool enough to preserve the fruits from the trees growing in the backyard, dark enough to keep the dried herbs that had been grown in the garden fresh in their labeled containers.

My grandparents on my father's side had moved to Detroit from the Louisiana bayou so Grandpapa could work at Ford Motor Company. In Louisiana, they had kept milk cows, grown strawberries, and raised their family in a house that was not much more than a tar paper shack at the end of a dirt road, bordered on both sides by swampland. They knew how to preserve their food, not out of novelty or political correctness, but out of necessity, incorporating all methods of food preservation to provide the family with a steady supply of nutrients. They had figured out how to bypass the prepackaged, costly, and distant world of the supermarket and still eat sun-dried tomatoes, dilly dough pickles, sauerkraut, and sour cherry pastries, even out of season.

To a child, the mystique and magic of this art did not include the stark realities that my grandparents lived with: poverty, hunger, and marital hardship. Instead, the ability to preserve one's food smacked of the autonomy and comfort associated with self-empowerment and hopeful inspiration.

Recently, I asked Dad how they ate fresh meat when he was a kid. He described how Grandmama would pull a piece of pork from a barrel stuffed full of meat that they had slaughtered and butchered and stored for weeks in the rendered pork fat that was poured to its brim. She would fry the pork for their lunch before they returned to work in the tobacco and strawberry fields.

Preservation arts such as these are being lost because there are few left to teach us how to use them and because our busy lifestyles seem to require modern conveniences such as refrigerators to store our food.

I want to be able to survive without a fridge and to maintain control of our food source, so at Fat Rooster Farm we are committed to growing, harvesting, and preserving our food. We are rewarded with tomatoes, broccoli, sweet corn, beets, Swiss chard, kielbasa, bacon, sun-dried tomatoes, parsley, pesto, potatoes, and parsnips—to name a few. Beginning in late March, when the sap begins to flow from the sleepy sugar maples and ending in late November, after the last of the livestock has been butchered and stored away in freezers, we use several preservation techniques to can, freeze, smoke, dry, and juice what we've grown and harvested. It is a challenge and an honor to have these skills, rooted in the necessity of knowing how to survive and fueled by the desire to live with respect for our planet's natural timekeeping of a sustainable harvest.

A Gilfeather turnip grows at Tunbridge Hill Farm in Tunbridge, Vermont. Either developed or discovered by John Gilfeather of Wardsboro, Vermont, in the late 1800s, the heirloom vegetable even has its own festival in the same Vermont town every year. Some say the vegetable is actually a rutabaga. In any case, it

the joy of keeping a
root cellar

Drops of early summer dew lie on a leaf of Red Russian kale at Tunbridge Hill Farm. The variety can be picked young for salads or picked in winter, when it's covered with snow, and eaten raw or cooked in soups.

" Savoring something—a spice, a radish, a piece of cheese—brings us back home to the world in which we walk and breathe. It slows us down. Taste is social. We come together, sit and talk together around food; we clink glasses and laugh and engage in small gossips and whispers in the presence of local beers or wines, tisanes and cakes with gooseberry preserves and clotted creams, or thin wafers bearing full fatted cheeses daubed with slices of purple figs. It is how we share being alive. We can engage in the virtual world of iPod music and TV drama, but there is no virtual world of taste. It is in our mouth, and every day our mouth connects us to place. "

—*Blessed Unrest*, Paul Hawken

Introduction

I HAVE A TEN-YEAR-OLD WHO has become seduced by the world of prepared food. Anything in tiny, prepackaged containers brings a gleam to his eye and an argument between us. No longer will he happily go out into the garden in search of a carrot to pull from the ground, sweetened by the clay in this Vermont soil, or grab the stalk of a Brussels sprout to pull the tiny blobs away and pop them into his mouth. He used to graze on the greens of onions and yank radicchio leaves from the rows to stuff eagerly into his mouth. Instead, he's more interested in the vitamins described on brightly colored wrappers, or cheese and crackers housed in little rectangular plastic containers, complete with a plastic paddle.

It's partly my fault, giving in to the peer pressure that he's confronted with. At summer camp the other kids stare at his lunch of homemade dill pickles, black cherry tomatoes, a peanut butter and black currant jelly sandwich (from jelly that his father has made), and homemade mozzarella cheese. I want him to feel like he fits in, so I acquiesce and buy the individual applesauces (still without sugar), the juice boxes (still 100 percent juice), the single strings of cheese (at least from Cabot). I have rationalized the choices of convenience and homogenization every step of the way.

Modern society carries with it a general perception that luxury and well-being are hand in hand with instant gratification and social hierarchical status. The idea that leisure does not include self-reliance has reduced our ability to care for ourselves. It is more likely that a well-stocked pantry is looked upon as a necessity for those without the ability to purchase the same products outright rather than an accomplishment to be praised or even envied.

Fast food chains like White Castle, McDonald's, and Burger King were invented in the 1940s and 1950s. Soon, these pit stops became quick food fixes for people on the go.

We are all relying more and more on big-box stores for groceries. Instead of visiting the butcher for meat and the greengrocer for veggies or reaching into the cupboard for tomatoes they put up themselves, most everyone I know hops into the car and heads for Costco or Stop & Shop.

Now there's nothing really *wrong* with shopping at stores—after all, we're all busy, and perhaps you try to buy organic or local whenever possible. And maybe you're trying to grow some of your own vegetables, which is great. If this is the case, consider seeking out local growers, or save and start plants from your own seeds, rather than visiting Lowe's or Kmart for a couple of seedling starts. If you depend on the mega gardening centers, you risk higher incidences of deadly food-borne illnesses due to consolidated packing and processing facilities, loss of community-based businesses, compartmentalization of commodities, even widespread distribution of plant- and animal-borne diseases.

There is good news, though. In the past few years, terms like "locavore," "sustainable," "renewable," "slow food," and "small farmer" have been tossed about almost as frequently as the stats for the failing stock market and the price fluctuations in gasoline and crude oil. The attention paid to First Lady Michelle Obama's garden at the White House rivals the publicity garnered by the Victory Gardens planted there by First Lady Eleanor Roosevelt. There is a movement afoot urging Americans to become more independent and at the same time more community oriented. Of course everyone won't—and can't—run a full-blown farming operation. Growing a few heirloom tomato plants for homemade spaghetti sauce is one thing, but picking and preserving the fruits from 600 plants is not realistic for most of us.

As Sandor Ellix Katz states in *Wild Fermentation*, "Not everyone can be a farmer. But that's not the only way to cultivate a connection to the Earth and buck the trend toward global market uniformity and standardization."

If you don't have land or the ability to grow your own vegetables and fruits or raise your own meat, milk, and eggs, be creative. Barter your tax-accounting skills for chicken and beef for the freezer; lend your carpentry skills to a farm in exchange for a bushel of cukes for pickling; or swap your housecleaning services for potatoes to cellar for the winter months. Use this book and its list of resources to learn how to preserve food by canning, drying, fermenting, freezing, root cellaring, and smoking. Enjoy the long, slow days of seasonal rest by eating what you have harvested and preserved.

Cutting off its stem and roots, the author harvests a Gilfeather turnip. It was the first season the heirloom rutabaga had been planted at the farm. "They're intimidating because they're so big," Megyesi said. "They can feed a family of eight."

PART ONE

Storing and Preserving Fruits, Herbs, and Vegetables

Macoun apples, and other varieties, store well and can be enjoyed throughout the winter.

" A well-planned garden will give a good variety of fresh vegetables and a plentiful supply for canning, freezing and storing. A few berry plants will assure seasonable fruit and perhaps some jelly and juice, while just one apple tree may provide apple pie year-round! A farm flock of sufficient size to provide three eggs a person a week, and a cow to produce the required minimum of milk, will materially help in providing a good table at low cost. If facilities make possible the raising of a couple of "porkers" or a calf, the amount of cash which must be spent for meat can be greatly reduced. "

—*A Practical Guide to Successful Farming*, Wallace Moreland, editor

1

Fruit, Vegetable, and Herb Harvesting and Preparation

Planning and Planting What to Grow

SEED CATALOG CHAOS SETS in at the farm just in time for our "down" time, when I'm no longer consumed with tending to weeds or harvesting vegetables, fruits, or crops to feed the animals during winter months. Sometimes my husband will try and hide the catalogs that arrive in the mail, as he knows what is

What's an Heirloom Variety of Fruit or Vegetable?

Typically, these varieties of fruits and vegetables have historical significance pertaining to early periods of agricultural practices. Fruits that have been propagated from grafts and cuttings as well as open-pollinated vegetables (meaning that seed saved from flowering plants will be true and capable of perpetuating the same variety) are typically not part of large-scale agriculture. Growers of heirloom varieties are concerned with the historical significance of these plants, or they are concerned with the shrinking gene pool of varieties. As an example of genetic loss, the number of cabbage varieties has dwindled by nearly 65 percent since 1981. Why is this important? First, if you're a home gardener just interested in saving some of your corn seeds to plant during the next season, unless you're saving seed from an open-pollinated variety, the future crop won't be the same as the previous season's. Second, not all corn varieties are susceptible to the same diseases; if all the corn varieties are lost, and only one variety is grown (a monoculture), the risk of losing the crop altogether to disease or pests becomes greater.

sure to follow. I have time on my hands, and poring over the flood of catalogs spells possible financial setback.

It is prudent to begin planning what you will grow before it's time to harvest: planning the kitchen garden should take into account what you'll need—and want—to eat, come winter. Choose the varieties of fruits and vegetables that best suit your tastes and soil conditions, and it will make storing them less complicated and more successful. For example, growing mounds of purple filet beans is rewarding in the summer, when they can be munched on raw, or sautéed and tossed with vinegar, olive oil, and kosher salt, but they're pretty disappointing if frozen or canned: Their brilliant purple changes to a dull green when they're heated. Fancy, sweet, white onions are perfect as slabs in between crusty French bread together with juicy, red tomatoes and smoky bacon, but they don't generally keep in the earth cellar for very long.

What Varieties to Keep and How

New varieties of fruits and vegetables are developed every year, and there's a definite movement to bring back or stabilize the dwindling varieties of heirloom vegetables. However, you may be disappointed to discover that some popular storage vegetables from the 1970s, such as the Ringmaster onion, may no longer be commercially available. *Garden Seed Inventory, Sixth Edition,* compiled by Whealy and Thuente (2004), is an excellent resource for growers interested in heirloom vegetable varieties. It lists all of the nonhybrid vegetable seeds available through small and large seed sources alike throughout the United States and Canada (see References).

Most seeds advertised in seed catalogs or for sale as individual packets come with descriptions indicating the best way in which to preserve them, be it fresh, frozen, canned, or cellared. Refer to these descriptions to plan what varieties will be available to enjoy fresh and which will be better keepers. Ask your neighbors what varieties they prefer, and why. If you plan to sell any of the harvest, ask your markets which they prefer.

When to Harvest Fruits and Vegetables

Use the following guide and key to help you determine when to harvest and how best to preserve your fruits and vegetables. Subsequent chapters will cover specific preservation methods.

Preservation Method Key

**Alcohol = A; Can = C; Dry = D;
Fermented or Pickled = LF; Freeze = F; Juice = J;
Oil = O; Salt = S; Smoke = SM;
Sugar (Preserves) = P; Root Cellar = RC**

American Mountain Ash (Sorbus americana)—D

In the fall, this native tree is often burgeoning with clusters of tiny berries that can be easily dried and stored indefinitely. In Maine, we would dry bunches of the berries over the woodstove and then use them like people use Airborne or Emergen-C today, as they are packed with vitamin C. Popping one or two at a time in your mouth is like sucking on a lemon. They are a little less bitter if you wait until after a frost or two before harvesting them. They'll store well in glass canning jars.

Apples—C, D, F, J, P, RC

Pick apples when they change to their mature color, be it red, yellow, or green. Early ripening varieties are not good candidates for long storage; use these to make juices, vinegars, and sauces, or dry them. Pick firm, unblemished fruits from the trees gently, leaving the stem intact.

Avoid storing apples next to cabbages; they will impart an off flavor to the cabbage. Apples also give off ethylene, a gas that accelerates ripening and sprouting and can cause premature spoilage of other vegetables.

One trick used by the French is to store dried apple rings in elderflowers so that they take on a pineapple-like flavor.

One of the biggest treats next to homemade applesauce is apple cider. Surprisingly, it freezes very well, and having it accompany a turkey dinner in the dead of winter is an unexpected joy.

Apricots—A, C, D, F, J, P

Once picked, apricots no longer develop in sweetness or flavor, so leave them on the tree until they have fully ripened to a rich golden color. If you've never experienced a tree-ripened apricot, you are in for a treat. My father nursed an apricot tree for years in chilly Vermont until it finally succumbed to our bitter winters. My two sisters and I would fight over the sparse fruits it produced every fall. I prefer canning the fruit to other methods of preservation.

. . . . Saving Seeds . . . . from the Harvest

The art of seed saving was born out of necessity, when the best-producing plants' seeds were gathered and saved by farmers to be used in the following season. Seeds were saved from open-pollinated varieties of vegetables and fruits, meaning from plant varieties whose seeds would "stay true" to the characteristics of their parents. While hybrids and cloned varieties of fruits and vegetables may produce higher yields, a more uniform crop, and fruits or vegetables that store or travel better than their open-pollinated counterparts, they cannot be maintained by saving the seed from the planted crop.

- In the late 1990s, technology was developed known as Genetic Use Restriction Technology (GURT), commonly referred to as terminator seed technology. Seeds produced from these plants are sterile or carry specific traits that require an application of a specially purchased chemical to generate seed that remains true to the parent stock. The big companies that hold the patents on these technologies have vowed not to introduce them commercially, but the fact that these seed companies possess the power to keep the individual gardener from saving his or her own seeds is a frightening one.

- Even entire nations have joined in the regulation of seed saving. In 2004, the Coalition Provisional Authority (CPA) representing the government of the United States made it prohibitory under Order 81 for Iraqi farmers to save their own seed; instead, they are forced to purchase hybrid seed from commercial seed manufacturers outside the country. Many of these farmers have saved their seed for generations, so an enormous diversity of plant genetic material is at risk of being lost.

- In Brazil and Pakistan, 80 percent of farmers still rely on seed they've saved from previous crops for future harvests, but in the United States and Canada, seed company consolidation, mobile lifestyles of families who change where they're living every few years, and corporate agricultural greed have all contributed to a loss of seed sources and varieties of vegetables and fruits.

- From 1984 to 1987, nearly 24 percent of small seed companies offering open-pollinated varieties of fruits and vegetables were bought up, consolidated, or lost. At the same time, the number of available varieties of vegetables and fruits plummeted; from 1981 to 2004, as many as 64 percent of open-pollinated varieties of certain vegetables vanished from retail. Just as disturbing is the fact that out of nearly 2,600 open-pollinated varieties available for retail now, more than 50 percent are available from just one source, while less than 10 percent of varieties are available from three different sources.

- Take the cardoon, for example, a member of the thistle family, and considered inferior to its close cousin, the globe artichoke. Cardoon seeds were saved by the ancient Greeks and Romans at least since the fourth century B.C.E. The vegetable is still common in the Mediterranean today but is difficult to find in the United States. Eight open-pollinated varieties are grown, but only one variety is offered by more than three seed sources, while five other varieties are each sold only at five different sources.

- Despite the corporate free-for-all of massive seed source consolidations, buyouts, and irreplaceable losses of so much genetic material, it seems as though the trend may have at least stabilized. The Seed Savers Exchange (see Resources) tracks the inventory of seed catalogs and nonhybrid varieties of seeds available in the United States and Canada. They found that after 1990, the total gains in seed sources began to surpass those being lost. They also showed that while available seed varieties have increased since this time, the older varieties are continuing to vanish at a constant rate.

- There are several sources of open-pollinated seeds listed in the resources section at the end of this book, but if you want to try saving seed yourself, and you're new at it, consider trying to save seed from an open-pollinated variety of tomato. They're an easy plant to begin with, and the saved seed will stay viable up to ten years if it is stored properly.

- Whether tomato varieties will cross with one another remains controversial in the seed- saving world, but just to be safe, choose a variety that is not a currant, potato-leaved (the vines have leaves shaped like those of a potato plant rather than the toothed leaves of most tomato plants), or double flowered.

- Pick only fully ripe fruit, and pull out the seeds from between the chambers of the tomato. I like to cut the tomato through the middle and then scoop the gel and seeds out, or squeeze the gel and seeds into the container. Use a container such as a Tupperware or an empty yogurt cup. Make sure there is enough gel in the container so the mixture has enough liquid in it that the seeds don't dry out. Loosely cover the container, and set it aside for 3–6 days, until a whitish-gray mold forms on the top of the liquid. The mold helps to loosen the seeds' gel-like coating that inhibits germination while they are inside the tomato. The container will also smell nearly as bad as rotting onions, so keep it in an area of the house where you aren't likely to smell it.

- After the mold has formed, and you can see bubbles trapped below its surface, the seeds are ready to cure. You can add water to the container, swish it around, and then pour off the mold and inferior seeds, or you can strain the mixture through a mesh strainer and rinse the seeds off vigorously. Dump the seeds out onto a paper towel or paper coffee filter and allow them to dry completely before sealing them up. Some seed savers frown on saving them on paper and advise dumping the seeds into a glass or ceramic dish and then stirring them until they're dry over the next few days. I find it just as easy to strip them from the paper and gently separate them with my fingers before planting, and the seeds dry more quickly on the paper.

Artichokes (Globe types) and Cardoons—C, F, LF, O, RC

Because they are perennial, most artichokes will not produce the edible portion of the plant, the flower, until the second year. In cold climates, the plants can be grown from seed and tricked into thinking that they've gone into their second year. We plant the seeds in a protected, heated greenhouse as early as we do the onions. When they've grown their third or fourth true leaves, we leave them at temperatures as low as 28 degrees Fahrenheit for at least 20 days, so plan your planting time accordingly. Look for varieties that are well suited for colder climates.

In warmer climates, it may take a full two years before your plants flower, and in order for them to become fully established, hold off picking the flowers the first year.

A central bud will develop on the plant first. After this is harvested, side buds will form, much in the same way they form on a

BELOW: As an experiment, the author grew artichokes at Fat Rooster Farm. "It's not a cash crop," Megyesi says, because the Northeastern climate is too cool for artichokes to grow consistently and in abundance.

broccoli plant. One plant can produce as many as six buds in our cold valley, but the plants can't make it over winter.

Harvest the buds when they are full before they begin to open or turn purple. Use sharp scissors or pruning shears and cut the stem about 3–5 inches below the bud (the stem can be eaten like a cardoon's).

You can also dig up the artichoke roots and store them in an earth cellar. They can be replanted and will reproduce that season.

Cardoon should be blanched for about 3 weeks before it is harvested, or the outer leaves become bitter. You can do this by banking soil around the base of the plant and tying the leaves closed, or by putting paper, cardboard, or straw around the base. Trimmed of their outer leaves, the heart and inner stalks can be stored in plastic up to 2 weeks. Mulching the plants in the garden will also help keep them longer for a later harvest.

Artichokes (Jerusalem or Sunchoke)—F, LF, RC

We use these plants as a natural border around our summer dining area; they easily reach a height of 10 feet tall at the end of summer, and bright yellow sunflower-like flowers crest each plant. In the fall, dig them as you would a potato. The crunchy tubers will stay well stored in the root cellar. Because they are prolific, I never worry about overharvesting and leaving enough for next year's crop. Waiting until a heavy frost will make the tubers sweeter.

Asparagus—C, F

One of the most difficult tasks for me to bear was to wait for 3 years until the asparagus patch was ready for picking. There they were, two 80-foot-long rows, with crowns reaching 6 feet into the sky, and stalks as big around as a man's thumb . . . it is best to resist the gluttony and let the roots develop in cold areas. In warmer areas, you can cheat and do a light harvest during the second year. Twist the spears off just below the surface of the soil (you can use a knife, but you run the risk of cutting other emerging spears if you're not careful). An established bed can be picked steadily for 6 to 8 weeks. Resist the urge to nibble on all but a few later in the season.

Apprentice Janet Van Zoren digs
up Jerusalem artichoke roots at Fat
Rooster Farm. Van Zoren was working
at the farm for her second season—
since leaving, she has been traveling
across Europe and working on farms.

Some sources encourage drying or even forcing asparagus. First, the thought of dried asparagus seems downright unappealing to me, particularly if canning or freezing is an option. Second, the yield from forced asparagus is very minimal, and doesn't seem worth the effort. Still, if you need a spring fix, established roots can be thinned, planted into a bucket or box, and left to experience cold for several weeks. Then, bring the plants into an area that is room temperature, and the shoots will begin to appear. Harvest them as you would the established garden crop.

Basil—D, F, LF, O

Properly dried basil will retain its color and flavor for months. Freezing basil also works well, as does incorporating this versatile herb in vinegar or oil. Pick only blemish-free shoots to encourage continued growth through the first frost. Basil begins to wane when temperatures hover in the 50s Fahrenheit, but it can still be suitable for pesto.

Beans, Dried—C, D, F, RC

There are endless numbers of dry bean varieties. We usually grow these every other year, storing enough for the following two winters. Wait until the pods are brown and very brittle, and the beans inside have hardened and can't be dented. Field drying the beans is ideal, but don't let them mold in wet weather. Instead, harvest the entire plants and lay them out in a dry, protected area to finish curing. Attics are great for this.

Beans, Fava—C, F, D

This versatile bean can be eaten young, like a snap bean, or after it is mature. Harvest young beans when they are plump, but the beans haven't formed. The mature beans can also be eaten after they've formed in the bean pod.

Beans, Pole—C, F, D

Pole beans are my favorite; they look artistic climbing up rustic teepees made of sapling wood, and they are tasty at so many stages.

They're also quite diverse, ranging from runner beans to lima bean varieties, or even flageolet beans. Harvest when the seeds are still small and tender and not well formed in the pods. Snap them off just below their stems, and you will be able to pick again later in the season. If they are to be stored after drying, wait until the pods are dry and the beans have a hard outer shell.

Beans, Bush, Snap, and Filet—C, F, LF, D

These varieties are ready for harvest about 2 weeks after the plants flower. The girth and length at which to harvest will vary depending on variety, but beans should not be well formed in the pods for fresh eating. The seeds can also be harvested after they've dried and used like dry bean varieties.

Blackberries—C, D, F, J, P

Wait until the fruits have completely ripened on the vines for the sweetest berries. They should easily pull from the stems, and fruits should be dark black in color. I like to wait until the sun is up high enough in the morning to dry off the berries and drive away the mosquitoes that lurk in the cool of the bushes. They should be processed soon after picking, or you'll end up with blackberry wine on your hands.

Blueberries—C, D, F, P

The only way to make sure that the berries are ripe is to sample a few. The fruits should be almost blackish-purplish blue, very plump,

and easily plucked from the bush. I prefer highbush varieties, only because they're easier to accumulate and not as hard on the back. Blueberries are one of the most heavily sprayed crops, so if you don't grow your own, seek out organic berries for the least amount of pesticide and herbicide residue.

Beets—C, LF, F, RC

Beets are extremely versatile: Their greens can be sautéed when young and tender and taste somewhere between spinach and Swiss chard, while the beets themselves can be harvested small and pickled or grown until they're two or three inches in diameter and then stored in the earth cellar. If you're storing them for a long period of time, choose a variety specifically noteworthy for its keeping qualities. Pull the beets by hand so the roots aren't damaged; trim the tops about ¼ inch from the root.

LEFT: **Canned dilly beans in the root cellar at Earthwise Farm and Forest in Bethel, Vermont.**

Belgian Endive, or Witloof—RC

Not to be confused with endive, radicchio, or escarole, which are all chicories, this plant is grown for its root in the summer months and then later forced in the root cellar after all the other leafy greens have disappeared. Harvest the roots in early fall (the root takes at least 115 days to reach maturity) and store it in a cool place. After it is taken out of storage and kept in a warm, dark, moist place, it will form a "chicon" that is similar in taste to radicchio.

Broccoli—D, LF, F

Believe it or not, broccoli, brussels sprouts, cabbage, cauliflower, kai-lan, kale, and kohlrabi are all the same plant, *Brassica oleracea*, all of which have been cultivated to grow a different part of the plant. The edible part of broccoli is actually its flower. Harvest the dark green central head, preferably in the cool of the morning or evening, before the tiny buds flower and turn yellow. After the central head is harvested, several side shoots will continue to develop throughout the season. Plant broccoli in early spring for early summer harvest or late summer for fall harvest. Broccoli tolerates mild frosts, but will eventually succumb when winter arrives.

Broccoli Raab (Rapini)—F

Best fresh out of the garden, broccoli raab is more closely related to the turnips. Harvest the whole plant at its base when the flower buds develop and before the flowers open. The yield can be extended by snipping just the buds from the plants.

RIGHT: The author checks the red cabbage crop at Fat Rooster Farm. While only about 5 percent of the plants sell in summertime, the farm harvests the rest, wraps them in newspaper, and stores them for sale at winter market. Red cabbage is popular for dishes like sauerkraut and stir-fries.

Brussels Sprouts—D, LF, F, RC

This is a very fun vegetable to bring to farmers' market. Almost no one knows that the little cabbagelike vegetables grow densely along a stout stalk. Some people advocate pulling the lower leaves and topping the plants when the heads begin to form to encourage growth of the sprouts. I've done both and can't really see much of a difference. Make sure to pick the sprouts before they're yellowed; the lower sprouts will mature first. I like to wait until after a few good frosts before harvesting them—they're less bitter. Pickled brussels sprouts is one of my family's favorites.

Cabbage, Green or Red—D, LF, J, RC

Harvest cabbages that have firm heads and shiny leaves. Cut them at the root and peel away the outermost leaves (don't take too many

off). My favorite ways of storing cabbage are in sauerkraut form or in the root cellar, although you can also dry it.

Cabbage, Chinese (also Bok Choi, Pak Choi, Choi Joi, and others)—LF, RC

For storage, the whole plant, roots and all, should be harvested. Chinese cabbage does better in cooler temperatures, so plan on an early spring or fall harvest for tight, crunchy heads.

Carrots—C, D, LF, F, J, RC

How you plan to preserve carrots will determine which variety to grow; they range in size and shape from short and stubby to long and thin. Believe it or not, carrots were historically grown for their aromatic herb properties, so the roots in the eighth century were small and bitter tasting. Carrots weren't even orange until the Dutch began experimenting with their cultivation in the seventeenth century; the ancient carrots of the Middle East and Central Asia prior to this were reddish purple or yellowish.

Carrots are ready to harvest when their foliage is full and dark green and their flesh has developed its full color. Most cultivars can simply be pulled from the soil by grasping the leaves close to the root, but when they are grown in heavy soils such as clay, a trowel or fork may be needed to loosen them.

When planting, be sure to properly thin carrots so that they're given enough room to grow, or they'll never reach mature size.

Cauliflower—D, LF, F

Unless you grow a colored variety of cauliflower, such as Cheddar or Graffiti, cauliflower needs to be blanched or it will develop a yellow or otherwise discolored look that doesn't ruin the vegetable, but doesn't make it look as appealing as the traditional creamy, white heads are. Blanching is easy: Simply pluck one or two of the leaves and lay them on top of the head, or tie the leaves around the head to block out the sun. The heads are ready to harvest about 1 week after this treatment.

Celeriac—LF, RC

Also known as celery root, this vegetable has been widely used in Europe since the seventeenth century, but is completely underutilized in North America. A cousin of celery and Hamburg parsley root, it has celery-like leaves and a large, bulbous root that resembles a turnip. They're best started in early spring and planted as soon as the ground is workable; harvest the roots when they begin pushing out of the soil. There are several smaller roots that should be trimmed off before eating, but leave them on if you're planning longtime storage.

Celery—D, J, LF, RC

Stalks can be blanched by mounding soil, hay, or straw around the base at least 2 weeks before harvest. The whole plant can be dug up in the fall and brought into the root cellar for a winter's worth of more slender stalks, perfect for flavoring soups and sautés.

BELOW: Morning light reaches a row of celeriac at Tunbridge Hill Farm. While it's part of the celery family, celeriac is not as fibrous as celery and is ideal for grating into soups and stews. The root vegetable is also a good storage crop for winter.

Cherries—A, C, D, F, J, P,

Both sweet and sour cherry varieties should be allowed to fully ripen on the trees, but if you're planning to use them for baking or preserving, and you're having a race with the birds to see who gets the most fruit, sour cherries can be harvested on the under-ripe side. The fruits should be shiny and red, but not split open. Eat or process the food within 2 days of harvest; sour cherries are particularly fragile and will bruise and turn brown if not used directly after harvesting.

Nanking cherries are native to China and have shrubby growth that can reach a height of 20 feet. The fruits are smaller than sour or sweet cherries, but make excellent jams, jellies, and syrups. They usually produce within their second or third year, are winter hardy, and are excellent hedgerows. Harvest the cherries when they are soft and fully ripe on the bush.

Chives—D, F, RC

Chives can be harvested throughout early spring to late fall. Cut the leaves back to the base of the plant, so subsequent harvests won't leave you with bunches that have browned tips. Don't overlook the stunning light purple flowers as food; harvest them (discard the woody, green stem of the flower) before the heads open fully. You can't really store chives in the root cellar, but they are very easy to keep inside on a sunny windowsill. In the late fall, after several frosts, dig up a clump and put them in a roomy, well-draining pot with fertile potting soil. Keep the leaves snipped back to stimulate the plant to keep producing. Transplant the chives back into the garden after the soil becomes workable in the spring.

Cilantro—D, F

As for chives, cut the leaves back to the base for multiple cuttings; eventually, the plant will go to seed, and the seeds are known as coriander. Replant seeds every 2 or 3 weeks for a continuous supply. To harvest coriander, let the seed heads dry out completely before harvest, then pluck the seeds from the heads and store them in an airtight container.

Collards and Kale—F, RC

Both plants are extremely cold hardy; in our garden, even after hard 20 degree Fahrenheit frosts, the plants will revive when the sun comes out and the temperatures warm back up into the 30s. Pick the leaves from the plants, rather than cutting them off squarely (they eventually look like the truffula trees described in Dr. Seuss stories, with long, thick stalks with leaves jutting from the tops). Whole plants can be harvested and kept in the root cellar for up to 3 weeks, as long as humidity remains at 90–95 percent, in plastic bags (poked with holes) or draped with wet towels.

Corn—C, D, LF, F

Corn is ripe when the skunks and raccoons start raiding the garden for it here at Fat Rooster Farm. Another indication is when the silks turn dry and dark. I usually pull back the husks to look at the tips to see if they are filled out and plump. Grasp the stalk with one hand and snap the ear backward away from it. Old-fashioned sweet corn usually becomes starchy if not eaten or processed shortly after picking; newer, hybrid varieties have a longer harvest window and storage capabilities.

Popcorn, flour, and ornamental corn should remain on the stalk until completely dried. The outer husks will be brown and parchment-like in texture.

Crabapples—P, RC

Harvest crabapples as you would apples. They are excellent keepers in the root cellar, suspended in burlap or onion bags with the same temperature and humidity that apples require.

Cranberries—D, F, J, RC

A native of North America, these fruits pack an excellent source of vitamin C. And contrary to popular belief, they don't grow in water; they require highly acidic soil with lots of peat mixed in. Harvest berries when they turn red, usually in the late fall. Plants in a 4' x 4' bed are capable of producing up to 5 pounds of fruit.

Cucumbers—LF, RC

At the height of the growing season, cucumber vines should be checked at least every other day. Depending on the heat and precipitation, a fruit can grow from the size of a thumb to the size of a fat hot dog within 48 hours. If you leave the fruits on the vines, they will stop producing, so pick the overripe ones to keep the vines healthy (pigs love them). Slicing cukes should be long and shiny green. Dull-colored cukes are often too seedy and punky tasting. Pickling cukes are crunchiest picked 2-5 inches long, but they can be preserved as "cornichons" if picked at ½ inch to an inch long. Process cukes intended for pickling right away.

An overabundance of cukes can be stored up to 3 weeks in a dark, cool (40–55 degrees Fahrenheit) root cellar at 80–90 percent humidity. In early fall, just before frost, the last of the cukes can be harvested and stored in this manner—an easy way to avoid the monotony of a winter's menu of root vegetables for a few more weeks.

Currants—D, F, J, P

Let the fruits ripen on the bush before harvest, though picking them slightly unripe will give higher pectin content to the jelly. Red currants are best frozen or jammed, but black currants are more versatile and are less tart. Bushes are highly prolific, fruiting in their second year. One bush can easily produce more than 10 pints of berries in one season, beginning in July.

Dill—D, F

Harvest leaves by cutting the plant to the base or by stripping individual leaves (cutting the entire plant will lengthen the time before it bolts to seed, but if you need seed for pickling, you may want the plant to mature and only strip it of its lower leaves). Dill is surprisingly frost tolerant; in a protected, unheated hoop house, it can withstand temperatures as low as the upper 20s Fahrenheit.

In seed catalogs, corn was historically referred to as Indian, dent, field, or sugar corn. Indian corn referred to the brightly colored ears with starchy kernels that were commonly ground into flour; dent corn with its shrunken-looking kernels could be eaten fresh, roasted when the kernels were more mature and at the "milk stage," or used to make flour, corn meal, and grits after it had dried. Field corn referred to varieties grown for animal feed, and sugar corn was the precursor of modern-day sweet corn. Sugar corn was probably a mutation of dent or flint corn (or both). Many corn varieties were lost between 1930 and 1970, particularly sugar corn varieties, because the modern, sweeter hybrids are now preferred for fresh eating.

Sweet corn is categorized as follows:

Normal Sugary (su)—The old-fashioned corn flavor, and the only choice for growers interested in open-pollinated (nonhybrid) varieties of corn. *The Garden Seed Inventory* lists more than 50 varieties of normal sugary sweet corn still available. This type of sweet corn quickly converts its sugars to starch after picking or if not harvested while it's ripe. It's best used for canning and freezing unless it is to be eaten right after picking.

Sugary Enhanced (se) or (se+)—These varieties are hybrids between an (su) parent and an (se) parent and produce more tender kernels and slightly sweeter flavor. They can also be harvested and stored a little longer, because the conversion from sugar to starch is slower. These varieties must be kept isolated from supersweet (sh2) and dry corn and popcorn or they will become starchy and tough. You can isolate varieties by planting early, midseason, and late varieties, so crosspollination does not occur.

Super Sweet (sh2)—These refer to "shrunken kernel" varieties, and this gene adds extra sweetness and tenderness to the kernels. These varieties can be harvested and stored safely for up to 5–7 days after harvest. Keep them isolated from all other corn types.

Synergistic or Triple Sweet—The sweetest of sweet corn, these have 25 percent (sh2) kernels and 75 percent (se) kernels. The ears have the tenderness associated with the (se) varieties and the added sugar content of the (sh2). The extended harvest window and storage capabilities make it an excellent market sweet corn. It must be kept isolated from (sh2), dry, and popcorn varieties.

While touring the hoophouse at Fat Rooster Farm, Randolph Elementary third-grader Chris Lamson, 9, tastes a dill plant.

Endive and Escarole—RC

Plants should be blanched prior to harvest by tying the leaves up with a rubber band, or by placing a cardboard disc or container such as a plant pot over the plant. Eating unblanched plants is not harmful; they are slightly more bitter and piquant than blanched plants. Harvest the whole plant and replant them closely together in sand in the root cellar. Be sure to remove the rubber bands from the leaves, or they may grow mushy.

Fennel Bulb—F, RC

Harvest the entire bulb at the base when it is mature and before a seed head forms. Cut the leaves off to form a fan around the bulb, and store in the root cellar for up to a week (unfortunately, the bulbs don't store well before drying out and losing their flavor). Don't wash the bulb until you're ready to use it.

Garlic—F, LF, D, RC

In mid to late July, harvest the flower stalks or "scapes" formed by hard-necked varieties. These are great added to slow cooker of lacto-fermenting vegetables or as additions to stir-fries. The bulb itself will be ready once the top leaves of the plant have turned brown. I test the garlic by pulling it, beginning in June. The bulbs should have readily demarcated cloves, with firm papery sheaths surrounding them. If the cloves have burst from their sheaths, they won't store well. Garlic needs to be cured before it is stored, but don't let it lay out in the sun as you would for onions. Let it dry for 3 or 4 hours and then store it in a cool place out of the weather for at least 2 weeks. Don't keep garlic in a humid root cellar; store it in a dark, cool closet at about 50 degrees Fahrenheit and 50–60 percent humidity.

Gooseberries—F, P

These spiny treats are easy to grow and make superb jelly. Allow the fruits to ripen on the bush; they'll be full and plump like a highbush blueberry when they're ready to harvest. The fruits are covered in small spines, so wear protective gloves when you harvest them.

Grapes—C, D, F, J, P, RC

One of the most versatile of fruits, you can do a lot with grapes. When they're ripe, they'll give off a heady aroma: pluck a few to be sure, but they should be sweet and firm, not squishy or bruised. Cut the clusters from the vine with a sharp knife or scissors, or the ripe fruits may fall from the clusters. Harvest them early when they are cool or after the sun has set. Fall-ripening varieties are best for root-cellar storage. Harvest unblemished leaves and pack them in salt for stuffing or for adding to pickled veggies: the tannins in the leaves make the vegetables retain their firmness and crunchiness. Wild grape leaves can be used as easily as cultivated varieties.

LEFT: **Hard-necked garlic cures in an old chicken coop at Fat Rooster Farm. The garlic is hung to dry out for 2 weeks, but can be eaten green when its outer skin is softer and it has a milder taste. The farm grows three varieties of garlic from seed they began using in 1992.**

Nahm Jeem Gratiem
(Thai Crystal Sauce)

This is such an easy sauce to make that it's strange that it isn't more of a mainstay in our condiment lineup. The kick from the red pepper flakes and the garlic's bite make it a great addition to many foods from roasted duck to baked acorn squash. Use up your less than perky garlic that has been stored for the winter and those red peppers that have been hanging from strings in your living room for too long.

- 8 cups white sugar
- 4 cups water (nonchlorinated)
- 2 cups white or rice (preferred) vinegar
- I cup minced garlic (around 16 heads)
- 2½ tablespoons kosher salt
- ½ cup red chili pepper flakes (use a food processor or herb grinder)

Bring all ingredients except the red pepper to a rolling boil in a heavy saucepan, stirring constantly. Cook until the sauce reduces to a thickish syrup that coats the back of a wooden spoon. Add the red chili pepper. Pour into sterilized pint jars. Check to make sure the lids have sealed; the sauce will keep for several months in the pantry; refrigerate once opened.
Makes about 5 pints.

ABOVE: While visiting Fat Rooster Farm, former apprentice Whitney Taylor of Wellsboro, Pennsylvania, helps to peel cloves of garlic for Nahm Jeem Gratiem (Thai Crystal Sauce), which the author will can for sale at market.

Grapefruits—F, J, P

As for oranges, lemons, and limes, harvest the fruits when they are juicy and mature. Fall-harvested fruits tend to last longer than those harvested in spring or summer.

Guavas—F, J, P

Mature guavas have whitish yellow or green skins. Their flesh will be juicy and, depending on the cultivar, distinctly yellow, white or red. If you are going to use them to make jelly, they'll contain a higher pectin content if picked slightly underripe.

Horseradish—D, LF, O, RC

These roots can be harvested in the spring or fall. The roots tend to have more of a kick in the early part of the year. Snapping the roots off without digging them completely out will leave you with enough to propagate plenty more during the upcoming growing season. You can also plant the tops of the harvested roots and usually propagate new roots. To have a fresh supply all year round, harvest roots before the ground freezes and pack them in damp sand in 30–40 degrees Fahrenheit at 90–95 percent humidity. Horseradish leaves can also be eaten, steamed like mustard or broccoli, both of which are close cousins.

Kiwis—F, P

R. H. Shumway offers a cultivar of kiwi that is hardy from zone 3 to 7. Their climbing habit makes them perfect for a natural trellis or hedge division in the garden. Most cultivars require a male and a female plant to produce fruit, which ripens in early fall. The fruits are ready when they are soft and fuzzy.

Kohlrabi—LF, RC

An underrated vegetable in North America, these versatile vegetables can be eaten raw or stir-fried. The base of the plant swells to form a round or oblong shape. Harvest anywhere from golf-ball to softball size, depending on how you plan to use them. The thin root is easily snipped; then discard the leaves and peel the vegetable before use.

Leeks—D, F, RC

While it's entirely possible to overwinter leeks throughout most of the United States, they're easy keepers in the root cellar, so you'll be guaranteed a supply without battling the snows and cold. For milder flavored leeks, hill up soil or straw around the plants to blanch the lower part of the stem several weeks before harvest. Leeks are fall-hardy vegetables, so leave them in the garden until the ground threatens to freeze. Then, dig up the entire plant, leaving the soil clinging to the roots, and replant them in loose soil or sand in the root cellar. Most recipes call for just the white portion of the plant, but reserve the greener stem and leaves for soup stock.

Lemons—D, F, J, S, P

Don't harvest lemons from the tree until they're juicy. Lemon slices are easily frozen and later used in iced tea.

The author checks on lettuce growing in a cold frame in one of Fat Rooster Farm's hoophouses. In spite of near-zero temperatures outside, the soil in the frame was still soft and moist.

Lettuce—LF, RC

Too often we dismiss lettuce as only good as a raw green. Lettuce soup and lacto-fermented lettuce can be welcome changes, particularly in the fall, when the frost-hardy greens are still available, but you'd like a culinary change of scenery.

Harvest lettuce either by cutting it clean to the base (for a second or third harvest of baby lettuce or mesclun mix) or by harvesting the mature heads. It is best harvested early in the day, when it is cool and less prone to wilting.

Limes—see Lemons

For tart varieties, harvest as for lemons. Sweet limes are harvested in the winter. The flesh will be a light to dark green color when they are ripe.

Melons—F, J, P

Let melons ripen fully on the vine before harvest. They can be tricky to figure out. If you're new to growing melons, start out with varieties that are easy to determine when ripe, like cantaloupes or muskmelons.

Mulberries—F, J, P

As a kid, one of my favorite jobs was climbing the mulberry tree to harvest the sticky-sweet fruits. Ours was a wild tree, and as such, wildly unpredictable in its yield. Cultivars of today are far more reliable. Choose one that states this for best results. The fruits should ripen on the tree and, depending on which cultivar you choose, should be dark red, white, or even pink. Next to strawberries, these fruits are my favorite.

Mushrooms—D, C, F, SM

Most species of mushrooms are best preserved through drying and/or smoking, but some, such as chicken of the woods, are well suited to freezing. Be sure you are certain of the mushroom's identity, then pick only mature

Hongos Silvestres en
· · · · · · · · · · · · Escabeche · · · · · · · · · · · · ·
(Pickled Wild Mushrooms)

Mushrooms are very good dried, but pickling allows a much wider range of use, from serving as a condiment to adding to a salad or as a garnish for beef.

- 3½ pounds wild mushrooms, such as chanterelles, morels, or chicken of the woods
- 1/4 cup lemon juice
- 1 cup olive oil
- 2 cups white vinegar
- 2 teaspoons dried oregano
- 2 teaspoons dried dill
- 2 teaspoons dried basil
- 2 teaspoons kosher salt
- 1½ cups chopped carrots
- 1 cup finely chopped onion
- 4 cloves garlic, quartered
- Black peppercorns (about 4 for each half pint processed)

Use only very fresh, unblemished, properly identified wild mushrooms, free of insect damage or decay. Cut the stems about ¼ inch from the cap. Wash and then place in a pot with the lemon juice and water to cover. Bring to a boil and simmer for 5 minutes. Drain the mushrooms, discarding the liquid.

In a saucepan, mix the olive oil, vinegar, herbs, and salt. Add the carrots and onions and heat to boiling.

In each cleaned jar, place a garlic quarter and about 4 peppercorns. Use a basket or slotted spoon to fill the jars with the mushroom mixture. Top of the jars with the oil/vinegar liquid, leaving ½ inch headspace. Process in a hot-water bath for 20 minutes.

After the jar is opened, the mushroom pickles will keep in the refrigerator safely for one month. Yields about 5 half-pints.

mushrooms for storing. Consult a mushroom identification guide for precise instructions on identification and harvest, but basically, the mushrooms should be firm, not insect riddled or woody, faded in color, or shriveled.

Nettles—F, RC

An entirely underrated green, nettles are the first leafed plant that appears in our cold little valley, sometimes as early as mid-March. The stinging bristles are rendered completely useless after steaming, and the iron and vitamin C contained in the leaves can outscore spinach and oranges any day, especially in Vermont in March.

In spring, harvest only the tops of the plants (the first two or three whorls of leaves), wearing gloves for added protection. You can cut the plants to the ground, and they will regrow within 2 weeks for a new harvest. In fall, dig up a clump and transplant them to the root cellar in rich soil for a winter of blanched, vitamin-packed greens.

Okra—C, D, LF, F

I abhor okra in its natural state, but pickled it's transformed into a mid-winter treat, unlike cucumber pickles or dilly beans.

One of the biggest mistakes in harvesting okra is waiting too long until picking it. Make sure that the pods are no longer than 2-3 inches, or they'll be woody and too seedy. They're usually ready to harvest just a few days after the gorgeous, hibiscus-like flowers bloom. Process them immediately, or they will turn into leathery, inedible spears.

Onions—C, D, LF, F, RC

We begin harvesting our onions that we start from seed when they just begin to bulb and use them as scallions. Later, we harvest fresh onions the size of a silver dollar, greens and all, as bunching onions. The storage onions are left to mature until the tops turn brown and tip over. They're left in the diffuse sunlight to cure for 2 or 3 days before storing in a cool, protected area. After the tops have died back completely, they're snipped about ¼ inch above the top and stored in a cool, dry place.

LEFT: Redwing onions grown by Fat Rooster Farm are on display for sale at the farmers' market in Norwich, Vermont.

The smaller onions can be canned or pickled, while the larger bulbs can be stored into early spring.

Be sure to choose onion varieties that will perform well in your region; onions are biennial, meaning that during their first season, they spend their energy producing a large enough bulb to have reserve energy to produce seed in the second season. In the beginning of the growing season, the onion produces its top growth. As the solstice approaches, the bulbs begin to form, but whether the onion will develop a bulb is dependant upon the length of day. In the south, where days are shorter in the summer than they are in the north, long-day varieties will not produce bulbs.

Orach—D, F, LF

Also known as mountain spinach, this green is one of the oldest cultivated vegetables, grown by the Romans and still very popular in France. It's easy to grow and preserved much like spinach. Plant as soon as the ground can be worked. Harvest the whole plant by cutting it to its base to prevent bolting; several cuttings can be obtained throughout the growing season. The seed is easy to save and cultivate in subsequent growing seasons.

Oranges—C, D, F, J, P, RC

Depending on the cultivar, determining when an orange is ripe can be tricky. The fruits should ripen fully on the tree, but to be sure it's ready, cut one open and sample it. Fruits ripening in spring and fall tend to have greener skin than those maturing in the summer months. Whether or not the fruit can be plucked or cut from the tree is also dependant upon cultivar type.

Oregano—D, F

Trim off individual leaves, or give the whole plant a haircut. The woody stems should be separated from the leaves before drying or freezing. The flowers are also edible.

Parsley—D, F, RC

Pinch or cut off leaves from the main stem for drying and freezing; the entire plant can be dug up in the fall and planted in loose soil in the root cellar for greens throughout the winter, providing that there is an additional light source, like a lightbulb or basement window. Parsley can easily be kept alive in a pot on the kitchen windowsill and snipped as needed. The plant will produce seeds the following season if it is replanted in the garden.

Parsley Root or Hamburg Parsley—RC

This plant was first cultivated in Germany during the sixteenth century. It has never gained popularity in the United States (at one point, it was almost exclusively grown in New Jersey), but is now more available in seed catalogs. The leaves are edible, but it is grown for its long, parsnip-like roots that are eaten cooked in soups. The flavor is somewhere between that of parsley and celeriac or carrot.

Harvest mature roots and store as you would carrots or parsnips, making sure not to trim the tops too closely to the root itself.

Parsnips—F, D, RC

A spring favorite, these roots can be wintered in well-drained ground and dug anytime after the first hard frost in the fall. For added protection, mulch the row with straw or leaves or floating row cover.

Parsnips are stored in a container in the root cellar at True 4 Now Farm in Woodstock, Vermont. Potatoes, turnips, carrots, Jerusalem artichokes and eggs are also kept in the space.

Peas (English, Snap, and Snow)—F, C, D

The first year we grew peas, we put them right next to the leeks and onions. The plants had poor vegetation, and not one pod on them. Later, I learned that one of the cardinal rules of companion planting is never to plant any of the Allia (garlic, leeks, onions, and shallots) next to beans or peas, because they will inhibit the latter's growth.

Plant peas as soon as the ground can be worked (they don't tolerate heat well). You may also treat the seed with a bacterial inoculant, which will increase yield.

Pick English and snap peas when the pods are firm and plump, but don't wait until the peas are fully formed, or they won't be as sweet. Pick snow peas before they've filled out and are still lime green, rather than yellowish, in color.

Peas (Southern Types)—F, C, D

Also referred to as black-eyed peas, cowpeas, or field peas, these vegetables are more like green beans than their English and Asian

Five Things You Can
······Do with Winter Squash······
Gone Bad

All of the literature advises that winter squash needs a cool dry place to store, so the earth cellar, whose ideal humidity should hover between 90 and 95 percent, is not the solution for long-term storage. I've successfully kept squash in the earth cellar until the December holidays, but after that, mold and mush tend to take over. If you don't have a cool, dry room to store the pumpkins and winter squash in, here's what you can do:

1. *Throw the rotten squash to the pigs you're raising for meat.*
2. *Give them to the sheep—pumpkin and winter squash seeds are said to have antihelmenthic properties, meaning they control the infestation of internal parasites.*
3. *Throw them to the chickens. About midwinter here in snowy Vermont, the birds are bored out of their minds, so give them a squash treat to peck at.*
4. *Throw them on your neighbors' lawn. Okay, don't do this, it may get you in trouble. I have a neighbor who has an annual cantaloupe shoot, much like a skeet shooting match. Maybe you could do that . . .*
5. *Remove the bad parts of the squash, cut them in half, remove the seeds and stringy pulp, and roast them in a hot oven until you can stick a fork through the skin. Let them cool, and then roughly scoop out the flesh. It can then be frozen for later uses such as soups or pies. Believe it or not, it was historically more traditional to make pies and breads out of other types of winter squash than it was to use pumpkins.*

cousins, and can be eaten young as such or left to dry in the pods for soups and stews. Pods are long like green beans, between 6 and 12 inches at maturity. The mature pods can be harvested and dried under cover if the weather turns damp (prolonged moisture can cause the peas to mold inside their pods).

Peaches and Nectarines—F, C, D, J, S

Harvest fruit directly from the tree, when the flesh has turned the correct color (depending on cultivar), and it is sweet scented. Twisting the fruit from the branch can prevent bruising.

Pears—F, C, D, J, S, RC

Pick pears when the skin has turned lime–green, and when cut open, the seeds appear to be turning brown. The fruits will look full-sized. My sister has a wild tree in West Virginia that produces pears that keep in the root cellar until late February. If you leave them to ripen on the tree, the flesh will turn grainy. You can still harvest these, just don't freeze, can, or root-cellar them.

Peppers, Sweet—F, C, D, RC

Harvest before frost, either young, when still green, or after they have matured to ripe color (red, orange, chocolate, yellow, or purple, depending on cultivar). The more you harvest, the more fruits a plant will produce. Leave the stems on if you intend to store them in a root cellar in early fall, when temperatures don't go below 40 degrees Fahrenheit, and humidity is 80 to 90 percent.

Peppers, Hot—F, C, D, RC

Harvest the same as for sweet peppers, except when root cellaring. Harvest the entire plant and suspend from the roots for a supply of peppers lasting throughout the winter.

Persimmons—F, D, S

Fruits can be harvested from the tree as soon as they are ripe and firm, but tart varieties can be left through several frosts to sweeten. Drying persimmons also removes any tart or bitter flavor.

Pineapples—F, D

Originally from South America, pineapples are now grown in the southern United States and Hawaii as well as Mexico. Pineapples reproduce asexually, meaning they are propagated from already established plants. The plant takes 15–20 months to mature before fruiting. About 10 months after it flowers, the fruit can be harvested. The exact time of harvest is often tricky; look for hardened, flattened eyes with yellowish to orange color. Pineapples harvested before fully ripening tend to stay fresh longer, but they aren't as sweet.

Plums—F, C, D, J, S

As for most fruits, harvest slightly less ripe if you intend to can or use in preserves, when the skin is dusky with a white bloom and still firm. Otherwise, leave them on the tree until plump and sweet. Be sure to follow instructions for specific cultivars; European plums ripen differently than Japanese varieties.

Pomegranates or Chinese Apple—F, RC

As the lemon is in the New World, pomegranates are a constant throughout the Middle East, Spain, and Italy. The tree prefers semiarid conditions and is grown in California most commonly here in the United States.

Harvest after the fruit is tinted reddish or yellow in the fall. They are excellent keepers and will store well for up to four months in the root cellar. Don't store fruits that emit a powdery puff when pushed near the blossom end; they will be dry and inedible.

Potatoes—F, C, D, RC

For new potatoes, harvest tubers by "stealing" them from live plants about 2 weeks after they flower (tickle around the base of the plant, just below the soil, so as not to disrupt the plant).

For storage potatoes, wait at least 2 weeks after the vines have died back to harvest. Keep freshly harvested potatoes in the dark, dry them without washing, and store in temperatures above 40 degrees, or they will turn starches to sugars and become mushy.

Pumpkins and Winter Squash—F, C, D, RC

Fruits should ripen on the vines but should not be exposed to heavy frost. Cut the fruits from the vines to prevent the stem breaking away. Curing the squash in a heated area such as a hoop house or greenhouse for 2 weeks will greatly expand the storage life by toughening the outer skin. Don't handle the fruits by the stems to prevent breaking them off.

Quinces—F, S, RC

Ripen on the bush before harvest after the skin has turned yellowish. How you plan to preserve the fruits will determine when you harvest; wait until after frost for freezing or making jam, but harvest before the frost if root cellaring.

Radicchio—RC

Harvest heads by cutting at base. Heads should be firm and full. Many cultivars are difficult to form proper heads, but the leaves will still be edible mixed with other greens. Radicchio is extremely frost tolerant, and even heads left in the field after several frosts can be trimmed away to use the crisp, scarlet centers.

BELOW: **After harvest, New England pie pumpkins cure in the morning light in the barn at Fat Rooster Farm.**

Radish—LF, RC

Harvest radishes as soon as they've formed their appropriate bulbs; Easter Egg varieties are round, while French Breakfast are more cylindrical. Test them often—if you leave them too long, they become woody and dry. Radishes also don't like hot, dry weather, so spring and fall plantings produce better quality yields.

Don't underestimate the radish greens; they're very good sautéed in butter and garlic, as are the radishes themselves.

Radish, Japanese—LF, RC

These radishes are also known as "daikon." Plant them in summer for late fall or early winter harvest, when roots are between 2 and 3 inches wide. The plants can withstand moderate frost, and the roots tend to have less of a bite after frost. Trim tops to 1 inch long for storage.

Raspberries—F, C, D, J, S,

Wait until the berries are soft, sweet, and juicy before harvesting them from the bushes. Process berries straight away to avoid mold. Berries used for preserves can be slightly less ripe.

Rhubarb—F, C, D, J, S, RC

Rhubarb needs to be babied the first two years in order for the plants to develop strong root systems. Your rhubarb patch should also be well mulched and well fertilized with aged manure. Harvest the stalks between 9 and 20 inches for the most tender; never eat the leaves or roots, as they are poisonous. If you continually pull the flowers at the base, you can harvest rhubarb well into strawberry season and beyond. Avoid cutting the stalks if possible; twisting them at the base will prevent the knife from piercing and damaging emerging shoots.

Rutabagas—F, RC

Wait until several light frosts before harvesting these sweet roots. Trim leaves to within ½ inch long and don't wash them until you're about to use them.

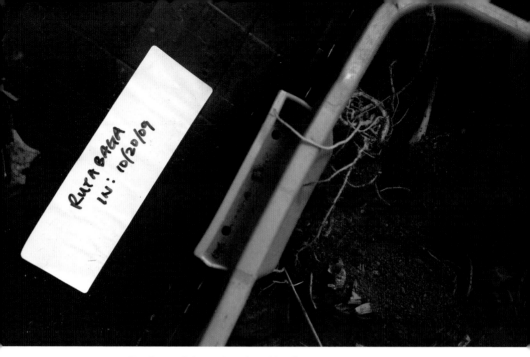

ABOVE: Rutabagas, left, are kept in a bin of moist leaves and celeriac is buried loosely in soil in the root cellar at Fat Rooster Farm. Set aside for sale at a later wintertime farmers' market, the vegetables store well in moist soil, leaves, sawdust, straw, or newspaper.

Salsify and Scorzonera—RC

Although these two vegetables are not closely related (they're part of the large Compositae family), they are planted, harvested, and used similarly. In Europe, this plant, grown for its roots, has been used since the thirteenth century; in the United States, Thomas Jefferson grew more salsify than any other vegetable. Its fragile nature at harvest and tendency to turn color if not dunked in acidulated water after processing have left it out of favor compared to parsnips or carrots (in 1981, eleven varieties of salsify seed were available for purchase, but by 2004, only three remained). Both are examples of vegetables that deserve our continued use (the wider the variety of root crops we have stored away, the less likely we'll become bored with our winter menu).

Harvest both roots after frost to make them mild and sweeter; harvest can also occur in early spring. Make sure not to store roots that have been bruised or cut off, and trim leaves so that at least ½ inch of vegetation remains. Don't wash them until ready for use.

Sorrel, French—F

Harvest the perennial leaves to add to soups and salads; their flavor is tangy and citrusy. Cut the plant back to its base to prevent bolting and woody leaves; new growth will occur in the fall. Planting in partial shade and moist areas also prevents bolting.

Soybeans—F, C, D

Grown in the United States as early as 1902, the edamame, or green soybean, explosion didn't occur until about 1998. Green soybeans are best harvested when the pods are firm and plump, but haven't turned color. Dry soybeans can also be harvested for soups when the pods are brown, but the stems are still green (you run the risk of the beans popping out of the pods and dropping to the ground if you wait too late).

Spinach—F, C, J, RC

Spinach is intolerant of heat; plant it as soon as the ground can be worked in the spring, or late summer/early fall for greens lasting well into the first months of winter. I have harvested spinach from

In late December, spinach grows in one of several cold frames at Fat Rooster Farm. Topped with two panes of glass, the box is in one of the farm's hoop houses.

the garden under 8 inches of snow that was crisp and firm as late as mid-December here in Vermont. Harvest individual leaves or cut the plant to its base and it will resprout.

Strawberries—F, C, D, J, P

Harvest fruits when they're fragrant and fully ripened. Check the plants daily to time the harvest and avoid fruit molding on the vine. Process the berries immediately, as bruising can occur when they are picked. (Unless you snip them carefully from the plant, bruising is almost unavoidable. However, I put up 50 pounds of berries a year, and I can't be bothered with the slow process of cutting them from the plant.) There are some cultivars that are day neutral and will produce berries all season into the fall. I find that the June bearers tend to be the firmest, sweetest berries, excellent for freezing and drying.

Sweet Potatoes—F, C, D, RC

Harvest the tubers right before or at the time of the first frosts of fall by gently digging them up and curing them for at least 10 days

BELOW: **Emma Hansen of Tunbridge, Vermont, searches for strawberries that are ready to pick while visiting 4 Corners Farm in Newbury, Vermont.**

Bright Lights Swiss chard is a hardy plant that grows at Fat Rooster Farm. Put into the ground as plants in mid-April, the chard continues to grow until it's covered in snow in mid-December.

in a shaded, dry area at about 90 degrees Fahrenheit and at least 80 percent humidity. After curing, they store best in 60 degrees Fahrenheit with low humidity—an upstairs closet is perfect to keep them all winter.

Swiss Chard—D, F, LF

Traditional harvest calls for trimming leaves when they're 4-8 inches tall, but it is easier to cut the entire plant back to its base and have it resprout in 2–3 weeks, depending upon the time of season. At Fat Rooster Farm, we set out transplants in the third week of April and have a continuous harvest of fresh chard through November.

I have a friend who brings in a chard plant and places it in a pot on her kitchen window. She keeps it going all winter for a tasty treat of baby chard leaves.

Table 1: **Average # Seeds Required and Yields Produced**

Vegetable	Number of Seeds**	
	Per 20' Row	Per 50' Row
Artichoke	10	25
Beans, Bush	160	400
Beans, Pole	80	200
Beans, Shell	160	400
Beets	260	600
Broccoli	34	85
Brussels Sprouts	40	100
Cabbage, Head	40	100
Cabbage, Chinese	60	150
Carrot	600	1500
Cauliflower	40	100
Corn, Sweet	40	100
Cucumber	120	300
Garlic (Cloves)	60 cloves	150 cloves
Greens, Mustard	300	750
Kale	90	225
Kohlrabi	300	750
Leeks	120	300
Lettuce (Head)	120	300
Melons	40	100
Onions, Bulbing	400	1000
Onions, Bunching	1000	2500
Onions (Sets)	60	150
Parsnips	280	700
Peas, Fresh	520	1300
Potatoes (Whole Seed Potatoes)	2 lbs	10 lbs
Pumpkin	27	68
Radish (Bunches)	720	1800
Radish (Daikon)	120	300
Rutabaga	240	600
Spinach	200	500
Squash, Patty Pan, Summer and Zucchini	60	150
Squash, Winter	40	100
Sweet Potato (Slips)	2	6
Swiss Chard	120	300
Tomato	10 for 7 plants	20 for 12 plants
Turnips	500	1250
Watermelon	40	100

** Number of Seeds, Unless Otherwise Noted

Yield	
Per 20' Row	**Per 50' Row**
50 chokes	125 chokes
16 lbs	40 lbs
30 lbs	75 lbs
3 lbs	6 lbs
8 lbs greens, 20 lbs roots	20 lbs greens, 50 lbs roots
15 lbs	35 lbs
12 lbs	30 lbs
12 heads	30 heads
12 heads	30 heads
20 lbs	50 lbs
12 heads	30 heads
1.5 doz	4 doz
24 lbs	60 lbs
60 hds (about 10 lbs)	150 hds (about 25 lbs)
20 lbs	50 lbs
15 lbs	40 lbs
20 lbs	50 lbs
30 stalks	75 stalks
20 heads	50 heads
20 fruits	50 fruits
20 lbs	50 lbs
20 lbs	50 lbs
60 bulbs	150 bulbs
15 lbs	38 lbs
4 lbs	10 lbs
20 lbs	100 lbs
60 lbs	150 l bs
20 bunches	50 bunches
20 roots	100 roots
30 lbs	75 lbs
8 lbs	20 lbs
40 lbs	100 lbs
20 lbs	50 lbs
8 lbs	24 lbs
15 lbs	40 lbs
60 lbs	150 lbs
20 lbs greens, 10 lbs roots	100 lbs greens, 25 lbs roots
14 fruits	35 fruits

Summer Squash and Zucchini—F, D, C, LF, RC

The smaller and younger these vegetables are, the fewer seeds they will contain; use the larger ones that have gotten away from you for stuffing or making breads and muffins. Pick 2- to 6-inch-long fruits to use in salads and pickles, and 8- to 10-inch fruits for sautéing. The flowers are also excellent on salads and for stuffing.

Thyme—F, D

Cut the woody branches back to the base to control the plant's growth, or harvest tips as needed.

Tomatillos and Husk Cherries—F, C, RC

Fruits are ready to harvest when they begin to erupt from their papery outer blankets, and the color changes to greenish yellow or purple (depending upon cultivar). Some varieties actually drop to the ground when they are ripe.

BELOW: **Even though the ground is frozen, a Gilfeather turnip continues to grow in the hoophouse at Fat Rooster Farm. Raised the first time at the farm, the heirloom rutabaga has been popular at market. Its sweet taste is a compliment mashed in with potatoes.**

Tomatoes—F, C, D, J, O, RC, S

While slicing and canning are the more conventional forms of pre-serving tomatoes, they are easily harvested and processed in a variety of ways. Harvest ripe fruits directly from the vine for the best flavor. Choose fruits that aren't overly ripe or squishy to avoid low-acid problems when canning.

When frost is imminent, harvest the entire plant and hang it by the roots in a dark place. Wrap the fruits in newspaper, and when a tomato is desired, pluck it from the vine, unwrap it, and bring it into the kitchen to sit next to apples. While they don't compare in taste to the garden fresh fruits of previous months, you'll still be able to enjoy tomatoes from the vine for up to three months.

Turnips—F, D, LF, RC

Harvest when the diameter of the root is 2-3 inches early in the sum-mer. The roots tend to become woody and bitter if harvested when the temperatures are hotter. Don't underestimate the greens of the plant, which will keep well in the refrigerator or root cellar in a plastic bag.

Watermelons—LF, F

Cracking the code as to when a watermelon is ripe is a tough one. Generally, there are three cues you should use to determine when the fruits are ready for harvest (don't harvest them from the vines until they are fully ripe). First, there is a small tendril close to the fruit on the vine, about 1½-2 inches in length. It should be brown and dry. Second, do the knocking test. The fruit should ring back a dull "poink" sound, as though it's hollow, not a higher pitched "pink," or metallic sound. Last, carefully turn the fruit over. The part lying in contact with the ground should be yellowish in color.

Zucchini—

See Summer Squash.

What's Available and When

A general knowledge of what can be grown and when it is ready in your area can better prepare you for when to think about preserving

Celery plants dug up from the garden were replanted in the dirt cellar floor at Fat Rooster Farm. Even with limited sunlight, the plants continue to grow.

foods. You might not grow or raise everything on your homestead, but you can plan to barter or you-pick from neighboring farms and markets. We grow enough strawberries for our family of three, plus three apprentices, but I always make a trip to Cedar Circle Farm in East Thetford to harvest at least 50 pounds of berries to get us through until the following June comfortably.

If you're unable to access the Internet and its astounding wealth of information, you're probably resourceful enough to seek out more traditional ways of sourcing out what's available and when. For instance, you might check with your USDA zone rating and compare it to conditions required for various vegetables and fruits. If you live in zone 4, you know that come May, morel mushrooms and asparagus, rhubarb and Jerusalem artichokes, ramps and nettles might be available. Go to the local extension agent for a list of farms in the area and what they produce. Visit your local vegetable and berry growers' association or organic farming association for a list of resources.

If you have access to the Internet, using the sources listed in the appendix will be a breeze. They are thorough and simple to use; they list month-by-month availability of products, specific farm names, and even farmers' markets in your area. The Internet is constantly changing, so searching by seasonal produce or locally available produce might turn up newer sites. Even the iPhone has an app called Locavore, which is GPS-based and can find local food anywhere you ask it to search.

How Much to Preserve

In the height of summer, after I've picked the thirtieth pound of bush beans, I become weary of them. I tend to lose perspective and forget to worry about our bare months. If there's anything that I can persuade you to do if you're confronted with a similar glut, it's to take baby steps and begin putting some of the food away for your consumption before the fall and winter. Here's how I do it. Every Saturday, after farmers' market, I look at the leftover produce. I prioritize it, according to family likes or dislikes, the ease in which it can be put away for future use, and whether it is worth more money fresh. Then I spend the next afternoon preserving those products.

For you, it may not be the excess harvested for market that spurs you to preserve for later. Take a walk around your garden in the evening with a clipboard. Jot down what needs to be processed, what needs to be weeded, what needs to be replanted. Sip a glass of wine or lemonade. Let the cat trail behind, wondering at your meanderings, and listen to the calls of the chuck-wills-widow out in the field (or maybe it's a house wren chortling from the neighbor's yard). Whatever it is, don't wait until fall to start thinking about putting food up.

How much to put by depends on your family's, friends', and community's needs. Think about the typical week. How many times do you eat green beans during the week? Once? Twice? How large is your family or community? Do you host potluck parties regularly or attend social gatherings that require food? Multiply this amount out over the course of time when there will be no fresh offerings (frozen or canned veggies don't hold a candle to the real deal from the garden, so don't go overboard!), and then add a couple extra servings for the unexpected occasion. Use table 1 to figure out how many seeds or plants you will need to achieve the proper yield.

A former bathroom in MaryKaye Maxwell's garage has been converted to a cold space for vegetable storage in Randolph, Vermont. Maxwell keeps the room temperature at about 50 degrees.

During a visit, Bradford Jones,
10, of Royalton, Vermont,
at far left, looks at the root
cellar dug by hand before
the house was built by Donna
Kausen and her late husband
Bradford, in Addison,
Maine from 1990 to 1993.

" As we shiver our way back to the warm kitchen with
parsnips in our pockets, a handful of potatoes, and a bag
of carrots, we feel very good about it all because we've
managed to grow it and keep it. "

—Root Cellaring, *Mike and Nancy Bubel*

② Cellaring Food for Storage

"I WANNA CALL DAD," *he says to me in a high pitch, after I turn on the tiny solar light, in the corner, near the east window, right where it's been for the past 20 years. He and Whitney are waiting at the front door until I turn it on. He's not afraid, he's full of wonder. He's amazed—just what I wanted.*

At first he protested when I said that I was yanking him from school and going Downeast to Maine, to visit my dear friends Donna, Geri, and Pete. This is the place where I learned how to dig my own well, to butcher a deer, to tan a hide, to can food, to cure garlic for seed, to smoke meat. It's where I learned how to eat a partridge just killed by a car, how to crack a lobster with only my hands, how to say good-bye to a friend dying of melanoma, a year and nine months after meeting him. It was a place that was grounding.

Right now, my ten-year-old, Bradford, who just 18 hours ago was in tears at the thought of not having Wi-Fi for 3 days, is just dragging out skins of cured skunks and Australian possums. Donna's house is phantasmal. "What's this one, Mom?" I think it's a mink.

Kyle likes to describe Donna as ornery. She's done everything, seen everything, and she truly could do anything else. She's buried someone who was her husband and her best friend; she knows how to shear sheep, rake blueberries, preserve any food you can think of, make the best homemade wine you could drink, tan any animal's hide, make friends with Feds and foes; she's truly an idol.

Geri and Donna have gone to dinner at Oscar's, a local lobsterman and friend, so we're alone in the house, giving me just enough time to snoop around with Whit and Bradford. I show them my favorite things: the hand-dug root cellar, right down to granite; the reading lamp made out of sheep's legs; the sepia picture of Bradford Kausen, whom our son is named after, looking contentedly out onto the water. Whit and my Bradford are rewardingly awed and amazed. This house has been completely constructed by this woman, her husband, and her friends' hands. She is dependant on her community, her friends, and loyalty to survive here in the Maine woods.

Perhaps the easiest method of preserving food for later consumption is the root cellar. Also sometimes referred to as an earth cellar, or underground storage, the simple principle behind root

cellaring is that it allows you to harvest food from your garden without processing and store it fresh for eating during the non-growing season.

As early as 40,000 years ago, native Australians discovered this method of preserving when they buried yams to preserve them for later. Walk-in root cellars were probably invented in England sometime during the seventeenth century, and the art was later carried over to North America by the colonists. Having a root cellar was just as important to early North Americans in the eastern parts of the country as a smokehouse was to the south.

Having a root cellar is now far more practical and versatile. It can be as elaborate

LEFT: Red Norland potatoes are stored in shelves built into the walls of a root cellar at Earthwise Farm and Forest. Farmer Carl Russell says he feels the cellar's stone walls regulate the air temperature better than poured concrete would.

as a specially designed, climate-regulated area that is insulated and contains a ventilation system, to a bin dug into the floor of an older home's dirt cellar or a barrel buried in a protected bank. Containers stacked on shelves behind the basement stairs, or an upstairs bedroom closet that stays cool during the winter months, can also be used to "cellar" your food. The basic principal that all these storage methods have in common is that they take advantage of the conditions that naturally occur during the fall and winter months. They provide a place to store food in a cool environment with regulated humidity and ventilation.

Root cellars can roughly be divided into those constructed above ground, those that are excavated, and those that are part of an existing structure.

An above ground cellar is made of insulatory material such as rock, concrete (poured or blocks), sod, or a combination of these. An insulated door leads into the root cellar. Disadvantages with this type of cellar are that temperature fluctuations are harder to avoid (they are more exposed), and any wood products used in the framing will eventually rot and need to be replaced because of the humid conditions within.

Cellars can be constructed by excavating into an existing hillside, lining the inside area with concrete or stone, and then constructing a front wall and entrance area. Advantages to this type of cellar are that three sides and the roof are far better insulated, though the time and costs of construction will be greater.

Existing structures can also be used to cellar foods. A storage space or bathroom in the barn or shop can be maintained to keep fall and winter crops during the months when it's too cold to use them comfortably. A bulkhead to the outside can provide access to an existing basement that has been insulated and lined with gravel or soil for cellaring food. We retrofitted an old sawdust bin that was standing idle in our barn with thick, reinforced concrete walls and an insulated door.

RIGHT: Joe Jones, left, and his father Bruce Jones of Jones Roofing and Masonry in Chelsea, Vermont. smooth fresh concrete poured for the roof of the new root cellar at Fat Rooster Farm. The vent at center is one of two used to regulate airflow and temperature in the cellar.

Many foods can be stored in the root cellar, from root crops to canned goods to salted meats or sides of curing beef and venison. Each food has micro conditions that must be met so that it's stored without spoiling. If you're just starting out using this preservation technique, it might be best to stick with the tried and true root-cellar vegetables first, like beets, carrots, potatoes, rutabagas, and celeriac. Brussels sprouts, celery, cabbages, and kohlrabi are all hardy keepers in the root cellar as well.

Controlling Temperature and Ventilation

Depending on the crop you store, the root cellar must maintain a steady, cool temperature with good airflow in and out (remember, these foods are still living tissue, and stagnant air will lead to mildew, rot, and spoilage). The object is to keep the temperature at the desired level for optimum storage, typically just above freezing to 40 degrees Fahrenheit (although most winter storage crops can tolerate temperatures ranging from 28 degrees Fahrenheit to 50 degrees Fahrenheit).

Six-year-old Bazel Russell walks onto the porch at Earthwise Farm and Forest. The black pipe in the center of the photograph leads to the farm's root cellar. Carl Russell said in the summertime, the afternoon sun heats up the pipe, drawing moisture from the cellar.

Within the root cellar the temperature will vary. Nearer the top of the cellar, the temperature is generally warmer, so crops that are more tolerant of warmer temperatures could be kept on shelves here, such as garlic and onions. Nearer to the bottom of the root cellar, store the hardier root vegetables, like beets, radishes, carrots, and celeriac.

Ventilation can be controlled simply by opening the covering of a buried box for a period of time or constructing vent pipes to your root cellar. Both intake and exhaust pipes are needed, where the shorter, higher pipe removes the warmer, lighter air, and the intake pipe brings in the colder, denser air. Our intake pipe is located on one side of our cellar, constructed of PVC pipe with a 6-inch-diameter opening. It stretches from the top of the building to just above the floor, so the cool air spills out of the pipe along the bottom. The top of the pipe opens to the outside air. The exhaust pipe is located on the opposite side of the cellar, leading to the outside air, but flush with the top of the cellar. If the pipe were to protrude into the cellar, it would create a condensation problem, where a dead space created between the dangling pipe and the ceiling would trap the warmer air exiting the cellar. Condensation dripping on the stored crops can lead to spoilage.

Controlling Humidity

Using a digital thermometer and hygrometer will measure both the temperature and humidity within the root cellar. These are inexpensive enough these days that having two—one at the top and one at the bottom, especially in larger root cellars—is a good idea.

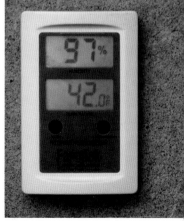

Improper humidity in the cellar can cause crops to shrivel up if too dry or become slimy and spoil if too moist.

As long as the humidity hovers around 90 percent, most crops will be fine. Exceptions are

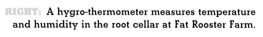

RIGHT: **A hygro-thermometer measures temperature and humidity in the root cellar at Fat Rooster Farm.**

Table 2: **Root Cellar Crops**

Root Cellar Crop	What to Plant	When to Plant	When to Harvest	
Beets	seeds	June-July	Fall, BF	
Broccoli	transplants	June-August	Fall, BF	
Brussels Sprouts	transplants	June-July	Fall, AHF	
Cabbage, Chinese	seeds	July	Late Fall, AMF	
Cabbage, Storage	transplants	June-July	Fall, BF	
Carrots	seeds	June (100 d prior to frost)	Fall, AMF	
Cauliflower	transplants	late June-July	Fall, BF	
Celeriac	transplants	June	Fall, AMF	
Celery	transplants	June	Fall, AMF	
Garlic	cloves	October-November	July	
Horseradish	roots	April-May	Late Fall, AHF	
Jerusalem Artichoke	tubers	April-May	Fall, AMF	
Kale	seeds	3 m prior to frost	Fall, AHF	
Kohlrabi	either	June-July	Fall, AMF	
Leeks	transplants	May (transplants)	Fall, AHF	
Onions	either	May (transplants)	July-Sept BF	
Parsnips	seeds	early June	Late fall/following spring	
Potatoes	tubers	early May	Fall, BF	
Radish, Winter	seeds	July-August	Fall, BF	
Rutabaga	seeds	July-August	Fall, AMF	
Squash, Winter/Pumpkin	either	May-June	Fall, BF	
Sweet Potato	slips	June	Fall, AMF	

Storage times listed refer to the best keeping quality and flavor results, along. They might be cosmetically challenged, but still edible.

* BF = before forst, AHF = after hard frost, AMF = after mild frost

Preferred Storage Temp (degrees F)	Preferred Humidity (%)	Storage Time
32–40	90–95	2 months up to 4 months
32–40	90–95	10–14 days
32–40	90–95	4–6 weeks
32–40	90–95	2–3 months
32–40	80–90	3–4 months
32–40	90–95	up to 6 months
32–40	90–95	up to 3 weeks
32–40	90–98	up to 6 months
32–40	90–95	up to 2 months
32–40	50–60	up to 9 months
30–40	90–95	up to 5 months
32–40	90–95	up to 6 weeks
32–40	90–95	up to 3 months
32–40	90–95	up to 4 months
32–40	90–95	up to 9 months
32–50	50–60	up to 9 months
32–40	90–95	up to 3 months
32–40	80–90	up to 6 months
32–40	90–95	up to 3 months
32–40	90–95	up to 4 months
50–60	60–70	up to 6 months
50–60	60–70	up to 6 months

though you can sometimes nurse stored fresh veggies and fruits garlic, onions, sweet potatoes, and winter squash, which like drier, not-so-cool temperatures. These crops are best kept in a part of the root cellar that is warmer, perhaps on a different level of the cellar, or in another part of the basement or house (see table 2).

The easiest way to keep high humidity is by having the floor of the cellar be dirt or sand, and periodically moistening it with water. Stored crops can also be loosely covered with moistened leaves, buried in moist sand or shavings, or covered with moist burlap bags. The walls can also be periodically sprinkled with water, or a plastic bucket with water can sit inside the cellar.

Controlling Light

With the exception of light fixtures that allow you to see your way around, the root cellar should be kept dark. Potatoes will sprout eyes, onions will begin to grow greens, and root crops will sprout leaves, depleting the roots of nutrition unless the cellar is maintained in darkness.

With an old door for a shelf, cabbage is tucked away in a cool, dark place in Donna Kausen's root cellar.

The new root cellar at Fat Rooster Farm is filled—with its regulated airflow and humidity, the cellar has extended the life of the harvest for the farm. It is an advantage to be able to take produce to sell at farmers' markets in the wintertime.

Shelving in the Root Cellar

Ideally, any object within the root cellar should be removable. Our root cellar has cinder blocks staggered to support roughly sawn pine boards that run up both sides of the cellar. With proper ventilation, the boards should be fine for at least five years, especially if they're removed each spring and properly cleaned and dried.

Don't use toxic materials within the root cellar.

Containers for Storage

Depending on who you talk to, the jury is out on whether wood has a place in the root cellar. Because it's prone to waterlogging and rot, wood doesn't hold up as well as plastic does. Still, if you're a purest, you may not want to store your vegetables in plastic. If you choose to use bushel baskets or apple boxes made of wood, be sure to remove them from the cellar in the summer and clean them carefully, looking for signs of decay.

Clam rollers store five
varieties of potatoes
Donna Kausen and her
two friends grow on
their property.

Do I need a permit to build a root cellar?

Always consult with your city or town officials to determine whether constructing a root cellar needs a permit. Obviously, burying a box of carrots packed with sawdust into a protected bank won't need the same consideration as constructing a new building. In some areas, the building might be considered an "agricultural shed,' while in others, digging a root cellar in the existing dirt floor of the house might be considered "unobserved construction" and illegal.

Onion bags and burlap bags are handy for onion, potato, garlic, and winter squash storage. Be sure to check periodically for condensation problems, and don't let water come in direct contact with the storage bags.

What Vegetables to Store in the Root Cellar

Below is a list of the more common vegetables that can be stored successfully in the root cellar. Once you've mastered these main characters, try branching out. Does your rhubarb need thinning out? How about your asparagus bed? Both of these roots can be

Built into a hillside at True 4 Now Farm in the root cellar is a cooler space than the ground-floor bathroom where farmers Lalita Karoli and Claude Richter had previously stored their crops. The root cellar cost the couple $3,000 to build from found, discounted, and custom-made materials in 2008.

brought in, buried in loose sand, and forced to create tender shoots in the spring. The same holds true for stinging nettles, a superb source of iron and vitamin C. Dig the plant up in the fall, after the stinging leaves have died back, and plant in the same way as you would the rhubarb or asparagus. Belgian endive, or witloof, a member of the chicory family, is grown by forcing the roots that have been planted the previous spring to grow tiny, edible chicons, pale green and tender.

Be sure to check your root cellar regularly. Remove any soft, shriveled, or moldy foods (some of these can be salvaged and fed

· · · · · ·Mustard-Glazed Brussels· · · · · ·
Sprouts

The brussels sprouts that I transplant to the root cellar take on an even sweeter, mellower flavor than the ones harvested in the field after the first few frosts. The mustard in this recipe gives the sprouts added depth, and glazing them creates an almost candied texture.

- *2 tablespoons salted butter*
- *2 tablespoons extra virgin olive oil*
- *1 pound small brussels sprouts, cut in half if over 1½ in diameter*
- *3 tablespoons pure maple syrup, grade medium amber or darker*
- *1 teaspoon Dijon mustard*

Melt the butter and oil in a large cast-iron skillet or other heavy skillet over medium heat. Wash the brussels sprouts, drain, and then add them to the fat to coat. Cover and cook for about 10 minutes. There should be enough liquid in the skillet to gently steam the sprouts; if the skillet becomes too dry add up to ¼ cup water. Remove the cover, then cook until all the liquid has evaporated and the sprouts are golden.

Stir in the maple syrup and mustard, coating the sprouts well. Cook for 1 or 2 minutes, stirring constantly. Season to taste with kosher salt and pepper.

Serves 4.

to the livestock, or trimmed and used for soup stock). Keep a log of when the foods were put in the cellar and how they stored. The following season, use these notes to improve your cellaring techniques. Table 2 gives a general description of when to plant, when to harvest, and how long to store root cellar crops.

Beets—When harvesting beets for storage, leave 1-2 inches of the tops so that you don't bruise the vegetables. Pack, unwashed, in loose soil or shavings kept moist.

Brussels sprouts—Plants can be kept right in the garden for an extraordinary amount of time. Here in Vermont, I regularly harvest brussels sprouts for Thanksgiving right out of the field. Let the plants experience several frosts before harvest; the frost intensifies their flavor. If you wish to cellar the plants, pull them up before the ground freezes, and plant them in moist soil. The plants will keep this way, providing crisp, juicy sprouts right into early spring. Alternatively, you can pick the sprouts and store them in a perforated bag. The sprouts will keep for up to 5 weeks if kept cold and humid.

Cabbage, Chinese—Wait until several mild frosts have occurred to improve the flavor of the cabbage. Harvest the whole plant, roots and all. Remove the most floppy outer leaves, then bury the cabbages closely together in loose sand. Water the roots well. Keep them as close to freezing as possible for the crispest heads. The outer leaves will probably wilt, but the inner leaves will stay firm for at least 3 months.

Cabbage, Storage—Use only fall-harvested cabbages that have firm, well-rounded heads. Pull the plants up by the roots, remove any loose outer leaves, and wrap individually with newspaper and store in a damp area just above freezing. Another method of storage is to tie them by the roots, suspended, or lay them loosely on a shelf. I find that packing them individually wrapped in newspaper or burlap in a plastic bin takes up less space in the root cellar.

Carrots—Harvest before the ground freezes. Cut the tops to about 1 inch above the root, brush off the loose soil, and store as for beets. Carrots grown in heavy soils like clay tend to be more fibrous than those grown in loose, sandy soil. Although not as sweet, the

· · · · · · · · · · Winter Root Slaw · · · · · · · · · ·

Spice up traditional slaw with celeriac and daikon radish to beat the winter cabbage and carrot blues. The different colors in this salad make it particularly festive.

Shred or grate the following vegetables:
- 2 cups red cabbage
- I cups carrots
- I cup celeriac
- I cup daikon radish

Combine and coat the prepared vegetables with:
- 2 tablespoons extra-virgin olive oil
- ¼ cup rice vinegar
- ¼ cup cider vinegar
- 2 teaspoons lime juice
- I tablespoon fish sauce (available at most Asian markets or specialty stores)
- ½ cup coarsely chopped cilantro
- ¼ cup maple syrup or 2 tablespoons sugar

Marinate the salad in the refrigerator for 30 minutes to an hour before serving.

Serves 8.

more fibrous carrots tend to keep longer. Carrots can also be over-wintered in the garden, although some may turn bitter.

Celeriac—One of the most underrated plants in the United States, this easy keeper is a sensational addition to the winter larder. Plants grown from mid- to late spring will tend to be less woody when harvested in the fall. Trim the tops from the plants to about 1½ inches. Dig the whole plant up and shake the loose soil from the roots; don't trim the roots too closely to keep from damaging the vegetable. Store in moist sand or sawdust close to freezing.

Celery plants dug up from the garden were replanted in the dirt cellar floor at Fat Rooster Farm. Even with limited sunlight, the plants continue to grow.

Celery—Harvest whole plants before the ground freezes. Cut the tops back to 3-4 inches, carefully avoiding the new center growth. Plant directly into the soil of an old cellar or in boxes of loose, moist sand. The plants will grow all winter, kept near freezing. With just a little light (from the basement foundation window or a lightbulb) the plants will continue to grow stalks that are naturally blanched. Avoid storing them near any of the cabbage family or turnips to avoid an off taste.

Garlic—Like onions, garlic must be cured before storing. After harvest, lay it in rows to dry, about 3–6 hours, then hang it in

In an old chicken coop at Fat Rooster Farm, hard-necked garlic dries before it is to be cleaned.

Mashed Potatoes with Celeriac

- *1½ pounds russet potatoes*
- *½ pound celeriac, washed, trimmed of long roots, and peeled*
- *¼ cup heavy cream*
- *3 tablespoons butter, or more, cut into slices*
- *Kosher salt and fresh-ground black pepper, to taste*

If the potatoes are relatively blemish-free, there's no need to peel them (there are lots of nutrients in a potato's skin). Cut them into quarters. Cut the celeriac into ½-inch dice. Put both vegetables in a heavy-bottomed pot with water nearly to the top of the vegetables. Cover and bring to a boil over high heat until the water comes to a simmer. Cook until the potatoes are tender enough to be pierced with a fork, about 25–30 minutes.

Drain and reserve the liquid. Mash the vegetables in a Chinois strainer, ricer, or potato masher.

Slowly heat the cream with the butter in a small saucepan. Stir in the milk mixture and as much of the cooking liquid needed to make the dish the consistency you desire. Season with salt and pepper and serve immediately.

bunches by its tops to dry in a cool, well-ventilated area out of direct sunlight. Garlic takes about 2 weeks to cure. After that, it can be braided or trimmed to about 1 inch above the roots, leaving at least 1 inch of stem on the bulb. After they've dried, clean the bulbs by removing the dry outermost skin and soil. Store loosely in burlap bags or onion sacks in a cool, dry (60–75 percent humidity), well-ventilated place. Softneck varieties tend to keep longer than hardneck, though both will keep until early spring if stored properly.

Kohlrabi—Store roots from traditional varieties that are planted during summer months; those planted in the spring tend to be woody. Harvest plants when they've reached about 2-3 inches in diameter. Remove the leaves and the root and store in moist leaves, sand, or sawdust.

Storage varieties like Kossack can be harvested up to 8 inches in diameter and stored for several months.

Leeks—In many areas, leeks can remain in the garden all winter, yielding new bulbs in the early spring. Whole plants can also be dug and replanted in loose soil in root cellar containers for harvest all winter.

Onions—When the tops have fallen over or have begun to turn brown, pull the onions and lay them, roots up, outside to dry for at least a day. Hang the onions as for garlic, or place on a drying rack until the neck is no longer loose and full of moisture and the bulbs have dried and formed an outer papery layer. Cut the tops to about 1 inch above the bulbs, trim the roots, and store in onion bags or burlap bags in a cool, dry place.

BELOW: **Harvested buttercup squash is set aside in the hoophouse to cure at Fat Rooster Farm. Curing gives the squash time for its shell to harden, making it resistant to disease and mold.**

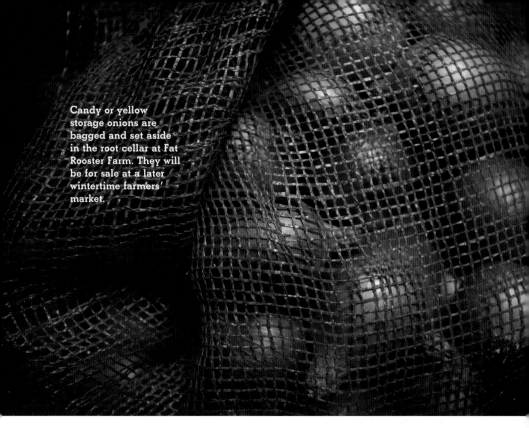

Parsnips—Store as for beets and carrots.

Potatoes—After the vines have died, leave the tubers in the ground for about 2 weeks to allow the skins to toughen (new potatoes that are stolen by digging around the base of the plant earlier in the season don't store well, nor do very small potatoes harvested later in the season). Cure the potatoes in a dry, protected area for at least 2 weeks. Store them in a dark location 36–40 degrees Fahrenheit. Potatoes stored at temperatures below this can turn their starches into sugar, making them overly sweet. If they're warmed to room temperature for 1–2 weeks, the sugar will revert to starch again and be of better eating quality. Don't let potatoes freeze in storage. Potatoes stored properly will last right into May.

Pumpkins and Winter Squash—Harvest squash when the stems are dry and the skins can't be pierced easily with a thumbnail. Be careful not to pull the stems from the fruits, or they will not keep well (cut these up and freeze them).

With the exception of acorn squashes, all pumpkins and winter squashes must be cured before storage. If frost is not imminent,

Butternut squash is left in the hoophouse to cure at Fat Rooster Farm. During an unusually cold, wet summer, the squash did well at Fat Rooster because the plants were surrounded by biodegradable black plastic that trapped warmth in the soil.

and warm, dry weather is forecast, they can remain in the field for the 2 weeks needed to cure. Otherwise, store them in a warm area where temperatures are at 75–80 degrees Fahrenheit (a south-facing porch is perfect).

Most root cellars are too cool and humid to store winter squash longer than 2 to 3 months' time. Some people store their squash in warm attics or heated basements; others put them in spare bedrooms or under beds for storage. Be sure to examine them carefully at least weekly and discard or use any squash showing signs of decay.

Because the skin of pumpkins is usually thinner than that of most winter squashes, it doesn't store as well. However, it can be cut, roasted and frozen for an extended period of use.

Acorn squash should not be cured and should be kept at cooler temperatures, about 40–50 degrees Fahrenheit for best quality.

Radish—Store only winter radish or daikon. Trim off tops to about 1 inch above the root and store as for beets, carrots, and parsnips. Although the flesh is more fibrous than the radishes of summer, their crunch adds a pleasing texture to winter salads and slaws.

Rutabaga—Cousin to the turnip, these sweet roots can be kept up to 4 months in the cellar, provided they are kept moist. I bury them in moistened leaves, but you can also store them like beets, carrots, or parsnips.

Trim the tops to 1½ inch, being careful not to damage the root itself. Harvest roots that are between 3 and 6 inches in diameter to avoid woody

RIGHT: **The author harvests a turnip for sale at market.**

insides. Harvesting after a few mild frosts often makes the roots sweeter.

Sweet Potato—Wait as late into the fall as possible to harvest the tubers to ensure the most growth. When the vines have wilted from frost, dig the tubers immediately, or the damage will travel down from the vine into the tuber and spoil it. Cure the tubers in a warm place (80 degrees Fahrenheit) to toughen the skins—behind the woodstove or in an empty hoop house is ideal. Store blemish-free tubers in a cool area (50-60 degrees Fahrenheit) with no more than 75 percent humidity for best results.

A rod stuck into an old bathroom vent controls the airflow to a cold space in MaryKaye Maxwell's garage. Keeping the room at about 50 degrees, Maxwell stores vegetables from her garden in the space.

What Fruits to Store in the Root Cellar

The root cellar is perfect for storing many types of fruits, particularly apples and pears. Even in the early fall, just before the first frosts, the root cellar can be used for short-term storage of watermelon, muskmelon, and cantaloupes, provided temperatures are cool enough. We've kept watermelon in the root cellar for up to 3 weeks.

Some fruits, like apples and crabapples, give off ethylene gas, which hastens ripening or sprouting. Unless you have a properly vented root cellar, keep apples away from other crops, particularly potatoes, so that premature ripening or sprouting will not occur.

Apples—For long-term storage (5-8 months), harvest mid- to late-fall ripening varieties. Avoid green or overripe fruits. Pick the fruit from the tree, leaving the stem intact to prevent tearing or bruising. Handle the fruits carefully, and store them at 32 degrees Fahrenheit with 80–85 percent humidity.

Apples that have fallen from the tree, called "drops," are less hardy in storage, but they shouldn't be overlooked. We've successfully cellared wild apples for 2 or 3 months that were not perfect; it just requires a more regimented inspection so that any soft apples can be removed before they spoil the rest of the stored crop.

Crabapples—Store as for apples, above.

Cranberries—In the Northeast, local cranberries can be found quite easily. I've kept cranberries in aerated plastic bags for up to 5 weeks. Store them at 80–85 percent humidity, between 35 and 40 degrees Fahrenheit for best results (cranberries freeze well for longer term storage).

Grapefruits, Oranges, and Tangerines—In our town, the local schools often have fundraisers around the holidays that feature sales of these fruits by the cases. They're easy keepers at 32–40 degrees Fahrenheit with 80–90 percent humidity. Inspect the fruits often for mold or soft spots (use these to make marmalades or glazes for baked meats). We've kept these citrus treats up to 2 months in the right conditions.

A pantry at Donna Kausen's home has bottles cemented into its end walls. Kausen said she was inspired by the same technique used in old miners' buildings in California. Staples like beans, dried berries, corn, grains, rice, and wheat are stored in the space.

Carboys holding fermenting fruit wines and mead at Donna Kausen's. She ages the wines in the carboys for about a year before they are bottled.

Pears—My sister in West Virginia discovered a wild pear tree in the woods near her farmstead. The pears are yellowish-green, rock-hard fruits, shaped like a Bartlett-type. In the fall, we harvest the pears by the bushel and store them in the cool root cellar between 32–40 degrees Fahrenheit at 80–85 percent humidity. They keep well past the holidays before we have to resort to canning the remaining fruits to avoid spoilage.

Pears should be picked when they're mature, but don't leave them to ripen on the tree, or their flesh will be grainy. Leave the stems on, and carefully harvest them to avoid bruises. For best storage, individually wrapping them is advised. To ripen them, simply bring the fruits into the kitchen and allow them to sit at room temperature (60–65 degrees Fahrenheit) for several days.

Using a Root Cellar to Store Other Foods

Beans, Dried—Dried beans store forever; I have black turtle beans that I've saved for 16 years. Every year, I plant some, save from that crop, and grow them again the next season. Make sure to store them in airtight containers, preferably with nonmetal lids, or store them in the dry part of the cellar, where moisture won't cause rust to form on the lids.

Butter and Cheese—The root cellar is a fine place to store cheese, yogurt, and butter on a short-term basis. Cream and milk that are to be used to make butter and cheese can also be kept up to 4 days in order to accumulate the proper amount required. All dairy products should be kept below 40 degrees Fahrenheit. Avoid freezing semisoft cheeses, yogurt, and milk, as the consistency will change after thawing. Hard cheeses and butter freeze well.

Dried Fruit—If kept in airtight containers, dried fruit will keep for several months in a cool, dry spot, perhaps in the same location your onions and garlic are stored. Avoid humid areas, or the fruit will become soggy, even moldy.

Eggs—Don't ever wash eggs intended for long-term storage—you'll remove the protective coating, called the "bloom," that keeps bacteria out of the porous shell. Eggs can be kept safely for 3-4 months in covered cartons at 33–40 degrees Fahrenheit and 70 percent humidity. As an extra precaution, float the eggs in a pot of cold water prior to use. Any that float to the surface should be discarded (perhaps a hairline crack was missed before storing them, allowing air and bacteria in through the shell).

Grains and Oats—Store as for dried fruits. If the grains have been freshly harvested, be sure that they've dried sufficiently before sealing them in airtight containers. You can do this by storing the grain in a burlap sack in a well-ventilated room for about 3 weeks, gently tipping the sack up and down periodically to dry the grains evenly.

There are several species of insects that have adapted their lifestyles to remaining dormant in grains until conditions are ideal for their emergence. This is why outdated cereals sometimes contain infestations of bugs; the eggs or larvae that were present when the grain was harvested have finally hatched out. You can either heat the grain in the oven for about 15 minutes at its lowest setting (130 to 170 degrees Fahrenheit), or you can freeze the grains below 0 degrees Fahrenheit for a week.

Meats—The fall root cellar is a perfect place to age fresh-killed meats, like venison, lamb, and beef. In our root cellar, there are two parallel galvanized pipes fitted with stainless steel rings to hold carcasses. The meat is kept below 40 degrees Fahrenheit while it is aging and then cut and wrapped 2–4 days after hanging. Fresh meat is extremely perishable, so it should be monitored closely.

Some people keep cured meats in the root cellar, but if it's perfect for root vegetable storage, it's probably too humid for meats like hams or dry-cured sausages. These products keep better in temperatures below 40 degrees Fahrenheit but at a humidity closer to 60 percent. The exception is pancetta, which needs to cure at 50–60 degrees Fahrenheit and 60–70 percent humidity; an early fall root cellar may be ideal in this case.

Nuts and Seeds—Store nuts in the shell until ready to use—they contain high amounts of natural oils that will turn rancid when exposed to warm temperatures and lots of oxygen. Whole nuts can be kept for up to a year in the root cellar at 32–40 degrees Fahrenheit and 60–70 percent humidity.

Store dried seeds in airtight containers in cold and dry conditions. Most root cellars that keep vegetables successfully will be too humid for nuts and seeds; keeping these foods with the onions and garlic will ensure longer storage.

Onions and garlic hang on the wall in Donna Kausen's bedroom. Kausen said the space is cool and dry enough to store allium vegetables there.

> **"**While they surely were unaware of vitamins, through intuition and experience our ancestors discovered that drying is suitable for only a few types of vegetables. On the other hand, it is perfectly effective for most fruits, mushrooms, certain herbs, and fish.**"**
> **—PRESERVING FOOD WITHOUT FREEZING OR CANNING**
> *The Gardeners & Farmers of Terre Vivante*

3

Drying Fruits and Vegetables for Storage

How Drying Works

INARGUABLY THE OLDEST method of food preservation, drying is also one of the simplest methods of putting food by and is used in almost every part of the world. While dried foods were essential for winter survival since perhaps prehistoric times, voyagers and travelers also regarded these foods as essential. Almost anything can be preserved through drying. In this chapter I focus on fruits, vegetables, and herbs; refer to part 2, chapters 8 and 9, for information on preserving eggs, meats, and dairy. Drying preserves foods by reducing the water content within it so that microorganisms and enzymes can't cause spoilage. In order for this technique to work, though, the food to be preserved needs to be impeccably perfect, not bruised or blemished, and cut into small pieces or thinly sliced, and you must work quickly. The air in which the food is dried must be dry and ideally warm. Sometimes, drying is combined with salting and smoking, especially with meats that are high in fat content and more prone to rancidity.

While many nutrients such as minerals are concentrated in dried foods, the process can also destroy valuable vitamins such as vitamin C and vitamin A, particularly in green vegetables. However, the vitamin contents of most dried fruits and tomatoes are far greater than those that are canned or frozen. How happy I was to have the previous season's dried tomatoes to crumble on salads when, the following year, our entire crop was lost (as were those of most other farmers in the Northeast) due to the big-box stores' unscrupulous marketing of plants infected with *Phytophthera infestans*, or late blight, to home gardeners. The infected plants sent windborne spores that spread like wildfire across the Northeast due to unusually damp and cool summer conditions. We generally can or freeze tomatoes, but drying a few as a backup proved essential.

Dried vegetables have limited uses, such as use in soups or broths, and their texture and taste often suffers after drying. Legumes, like beans and peas, generally do the best of the dried vegetables, retaining much of their nutrients and flavor. Herbs, too, are excellent dried, particularly in combination with one another and used as rubs and marinades.

Methods of Drying

Many foods require blanching prior to drying, so the enzymes in the food are deactivated, the living tissue is softened and prevented from drying out during storage, and the food's color is preserved. Blanching also cuts down on the time needed to rehydrate the food to its original condition. Simply plunging the food in boiling water for several minutes and then cooling it in ice-cold water prior

RIGHT: **Clockwise, from top, are the dried herbs—elderberry flowers, dill, and basil—that are among the variety sold at market by the author.**

to drying will work, but boiling tends to leach nutrients from the food. The preferred blanching technique is by steaming the food. This requires a little ingenuity if you don't have a steamer. You can use a large pot with a metal colander or wire basket placed on a rack inside. Have about 2-3 inches of water boiling in the pot, put the colander or basket inside, and then cover the pot for several minutes. Keep in mind that blanching by boiling will take about half the time it does to steam the same foods.

There are several methods of drying, from windy hilltops, to warm ovens fitted with drying racks, to commercially available dehydrators. Fruits and vegetables can be thinly sliced and strung together, just like popcorn or cranberry garlands are on the holiday tree; garlic and onions can be "dried" by fancifully braiding them together and hanging in a cool, dry place. The trick is to work quickly, to race against spoilage.

Using the Sun and Wind

The simplest method to dry foods is by utilizing the sun and the wind. All you need to do is remember to keep what you're drying away from precipitation, insects, and varmints (housecats are included in this category).

You'll need a drying tray, and this can be anything from a professionally made drying rack to a framed piece of hardware cloth (wire mesh), wooden slats, or cheesecloth placed on a well-ventilated surface, depending upon what the item is you are attempting to dry. One of the more common sizes for a drying rack or tray is 34 x 34 inches. Cheesecloth will fit across such a frame in one piece, and wooden slats or wire mesh can easily be fitted to such a shape. If you're using cheesecloth in the frame, be sure to use two strands of food-grade string from one corner to the other (to form a double X with a square in the middle of the frame) as support. I find hardware cloth or window screen is much easier to work with and easier to clean after the food is done drying. Be sure to turn the food often during the day, and only dry it during sunny, precipitation-free days. If you store it outside during the nighttime hours, you'll need to protect it with plastic or glass from the dew, insects, and cats.

Place the food in a single layer, skin-side down for produce, without crowding.

Using an Oven

Using the oven as a drier speeds up the drying process, reducing the risk of spoilage due to precipitation, insects, and varmints. The trick to drying in the oven is to make sure not to overcook the food. Another disadvantage to using your oven is that it can tie it up for hours. Many modern ovens won't ignite or operate at temperatures less than 170 degrees Fahrenheit, and ideally, the food should be dried at no higher temperature than 145 degrees Fahrenheit. Cracking the door and using an oven thermometer can help, but then you need to weigh the benefits of expending that much energy to dry your foods (usually the drying tends to occur during the warmer months, too, when you can't soak up the added heat from the oven to warm your house).

Use the oven racks themselves, spacing the food out evenly, skin-side down for produce, one piece at a time. If the food is too small to reach across the racks, use wire mesh, wire cooling racks, or cheesecloth draped across the oven racks (don't cause a fire!). Professional drying racks can also be purchased that are made to fit within the oven.

I don't consider a smokehouse to be an example of away to dry foods in the oven. With smokehouses, the added element of smoke curing changes the food's taste drastically. While drying removes water from food, it doesn't add to the flavor like smoking does.

Never use a microwave to dry your food. Some like to use it to blanch foods before drying, but you're just as likely to burn things to a crisp in the microwave if you attempt to dry them here.

Another method of drying not to attempt in the home is freeze-drying. While it is a superior method of drying (it preserves without loss of nutrients, color, or flavor), it requires complicated, expensive equipment. Unless you are planning to build your business around dried foods, it's too cumbersome for the average home gardener or small farmer to attempt.

Hanging Food to Dry

Some foods can simply be strung or bunched and hung and stored in a cool, dry area after harvest and preparation for drying. We often have lines strung from one end of the ceiling to the next, with hanging cobs of Indian corn and herbs waiting to be processed into flour or dried herbs.

Sriracha
(Rooster Sauce)

Use this sauce as you would any hot sauce, but be prepared for a tangy kick that can't be achieved using commercially available condiments. It's great with eggs, fish, and vegetable dishes, and is another great use off dried garlic cloves and chilis.

Fill a glass jar halfway full with peeled, rinsed garlic cloves. The rest of the way, fill with whole, stemmed but not seeded dry, red chilis, like Serrano, Thai bird, or cayenne. Add 1 tablespoon kosher salt per pint measure, then top the contents off with 5 percent vinegar. I use cider vinegar, but white vinegar is more traditional. Cover the jar and let it sit on a counter at room temperature. As the chilis rehydrate, top the jar off with vinegar. After 2 or 3 weeks, pour the jar's contents into a food processor and chop it until it's well blended. Keep in the refrigerator (it'll be good for months).

Using a Dehydrator

This is the method that I prefer for drying. It is easy and quick, efficiently drying foods like herbs in as little as 90 minutes. Our house is old, dusty, a haven for cluster flies in the fall, and the grounds are rich with skunks and raccoons that can easily acquire a taste for sliced tomatoes left out on a tray on the woodpile to dry in the sun. With a dehydrator, the temperature is regulated, and the food is contained within a protected shelf. While it consumes more energy than solar and wind power, a dehydrator could easily be hooked up

LEFT: Hardneck Rocambole garlic cures on fencing stretched across the rafters in the garage at Nicola Smith's home in Tunbridge, Vermont. Smith first got the seed garlic from a farmer friend four years ago and plants 60 to 100 cloves each year. Smith said she uses garlic in about 90 percent of the dishes she cooks.

to an inverter and battery that converts a solar panel's energy to AC current.

A good quality dehydrator can cost more than $100, but you'll gain back your initial investment quickly if you intend to do a significant amount of drying. I use the dehydrator to dry several pounds of herbs, which are then blended and sold as a value-added herb rub to be used on everything from poultry to stir-fry tofu.

Make sure that you choose a dehydrator that is made of materials that won't discolor the foods you're drying and are easily cleaned afterward.

Dehydrators tend to be noisy. If you don't have an out-of-the-way place to operate one, make sure that your model is quiet.

Be sure to buy a dehydrator with a thermostat on it. Different foods require different rates of drying, and without one, you risk burning the food.

The best dehydrators have fans that whisk away moisture and evenly distribute heat. This ensures that parts of the machine aren't hotter than others, leading to burned or underdried products.

Which Foods Should I Dry?

While many things can be dried, it's not necessarily the preferred method of preservation for every food. Certainly, your situation

should dictate the method of storage that you choose. If you're planning a hiking trip, and you'd like to munch on green beans, the preferred method of preservation, freezing, is not an option. Nor is canning them in a vinegar brine to produce vitamin C–rich dilly beans (imagine lugging that additional weight around). Maybe you can do without some foods for some parts of the year. Given the choice, I'd rather eat raw asparagus in season, doused with a little olive oil and vinegar and sprinkled with garlic chives, than eat reconstituted, dried asparagus in the winter. By then, the root-cellared celeriac or the Jerusalem artichokes can take the place of crunchy, bright asparagus until spring comes.

Storing Dried Foods

The key to storing dried foods is to avoid air and light. Both of these will cause the foods to become damp and soggy and promote the growth of mold. I prefer keeping the food in glass containers lined with paper bags to keep the light out. Large, commercial plastic food-grade containers, such as those for mayonnaise, are ideal for storing dried beans, peas, and herbs. Heavy-duty plastic freezer bags work, too, but keep them out of the light. Make sure to fill the containers to the top so that the food will be exposed to the least amount of air. Once you've dried them, it's best to store the foods

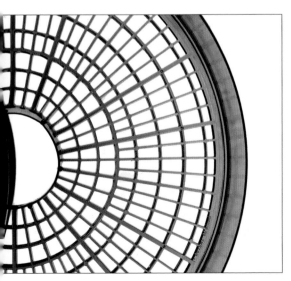

LEFT: Two types of dehydrators are shown; the version at left has a way to control the intensity of the heat. Tomatoes, for example, need more heat than plants like parsley.

RIGHT: One of the drying racks for a dehydrator shows how air circulates through the unit to dry plants.

at temperatures less than 50 degrees Fahrenheit. A dry corner in the root cellar (less than 50–60 percent humidity) is ideal—but for periods greater than 2 years, store the dried foods in the freezer. They'll take up far less space than freezing them fresh.

Using Dried Foods

Don't interchange measured amounts of dried foods in recipes calling for fresh ingredients; the recipes are taking into account the moisture content of fresh foods that is lacking in the dried.

For fruits and vegetables, you'll need at least twice and maybe four times more dried product to equal the same quantity of fresh ingredients. For herbs, use ½ teaspoon coarsely crumbled or ¼

teaspoon powdered herbs as a substitution for 1 teaspoon of fresh herbs.

Drying Fruits

Drying fruits such as apples, apricots, peaches, or bananas can cause them to brown due to oxidation. While it is commonly accepted for fruits like raisins, prunes, and figs to turn brown after drying, some eaters are put off when their banana chips are brown.

Commercially available fruits are usually treated with sulfur to avoid this, but using sulfur in the home requires special equipment and tedious precision.

The brown color won't affect the food's taste, but it won't look as pretty. An easy solution for the homestead is to use an acidulate (something that returns the vitamin C lost during the oxidation process that causes the browning). You can dip the fruit in either vitamin C tablets that have been crushed and added to water (use two 500 mg tablets to one quart of water), straight lemon juice (this tends to make the fruit quite tart), or a honey-lemon syrup (a ratio of one cup of water to one cup of honey and the juice of one lemon).

Fruits like berries, cherries, currants, and grapes can also be blanched to force their skins to crack and release their moisture more quickly. Don't blanch them for more than a couple of minutes in steam or half a minute in boiling water.

RIGHT: Parsley at left and sage are among the dried herbs bagged and sold at market by the author.

American Mountain Ash—Dry clusters of berries and remove the fruits after they've dried.

Apples—Use only firm, ripe, blemish-free fruits. You can pretreat them to prevent browning, but it's not necessary unless you are put off by the brown color. You don't need to peel wild or organically grown apples. Core the fruits and slice them thinly into rounds or wedges. The dried result will be slightly soft unless you dehydrate them at a high temperature to be eaten within a few weeks as chips.

Apricots—Remove the pits by cutting the fruits in half and then slicing thinly. Prevent browning if desired by pretreating them with a honey-lemon wash. They'll be slightly soft when dried. This is my favorite way to eat apricots.

Bananas—Peel the fruits and slice into rounds. To prevent browning, dip in honey-lemon wash. Dried fruits will be crisp when done.

Blackberries and Raspberries—Dry whole fruits until crisp (unless you're using a dehydrator, the berries may be somewhat chewier). The fruits are good dried when added to granolas, but I prefer frozen berries to dried ones.

Blueberries, Cherries, Cranberries, Currants, Grapes, and Mulberries—Quickly blanch stemless fruits in boiling water or steam to crack open their skins and release moisture. These fruits will have a leathery, "raisin" quality to them.

Gooseberries—Because they are typically thorny, drying is not recommended.

Grapefruits, Lemons, Limes, and Oranges—The peels of these fruits are dried rather than the fruits themselves (use them in recipes that call for zests for a bright burst of citrus flavor). If the fruits are purchased conventionally, they'll generally be coated with wax, so it should be removed by scrubbing the fruits with warm, soapy water. Rinse well and then peel away from the fruit. Scrape off the inner white pith and slice for drying. You can also grate the cleaned peels from the flesh to make zests. Be sure to grate only the peel, not the bitter white layer next to the fruit. Store dried zests in heavy plastic bags in the freezer.

Guavas—Because of their high moisture content, drying whole fruits is not recommended.

Kiwis—Cut fruits in halves or in thick slices. Pretreat with honey-lemon wash (page 103) if the fruit is intended for later use in cakes and breads. It can also be processed into excellent fruit leather.

Melons and Watermelons—Drying melons other than in a dehydrator is not recommended; their high moisture content keeps sun-drying and oven drying from being an efficient preservation method. Even dehydrators take up to 20 hours at 135 degrees Fahrenheit to completely dry. To prepare, remove the rind, cut in half, and chunk or slice into wedges.

Papayas—Split ripe fruits in half length-wise, seed, peel the outer skin, and slice thinly.

Peaches and Nectarines—To hasten drying, blanch quickly, remove the skins, pit, and slice thinly. You can pretreat them to prevent browning. The fruits will be leathery in texture and very sweet.

Pears—Peel the fruits, slice them in wedges or rings, and dip them in honey-lemon wash if desired to prevent browning. Dried pears will be soft and leathery in texture.

Persimmons—Choose firm, ripe fruits for drying. Slice them with a sharp knife ¼ thick. Dried persimmons lose their astringent taste.

Pineapple—Cut off the tops, peel the coarse skin, and core the fruit. Cut into rings or chunks (I love to add them to trail mixes) and dry them until the juice has disappeared (the fruit will be pliable and firm).

Plums—A dried plum is a prune. Cut the fruits in half, pit, and slice thinly. The slices will be completely black, like raisins, when dry, and are generally used similarly to raisins.

Pomegranates—Unless you're using these fruits for holiday decorations, drying is not recommended.

Quinces—An excellent choice for fruit leather.

Strawberries—Remove the stems from the berries and cut them in half. Fruits will be somewhat soft and chewy when dry.

Drying and Storing Herbs

Pick the herbs to be dried in the morning after the dew has dried, or at dusk, when the plants are cool. Try not to wash the herbs; they'll dry more evenly if you don't add moisture to them.

Dry herb types separately—parsley takes far less time to dry than basil or rosemary.

If you store the dried herbs whole, not crushed or powdered, they'll tend to retain their oils. Blends of different herbs are especially nice for cooking. Make sure when blending herbs to vary the amounts of each; the subtle tones of dill and parsley can quickly become overwhelmed if too much rosemary or sage is added. Lavender, mint, anise hyssop, epazote, and Mexican marigold should all be used sparingly in mixes. Dried lovage and celery leaf add salty qualities to herb mixes.

Drying Vegetables

Most vegetables require blanching or precooking in boiling water or steam for a short time prior to drying (see page 96). Exceptions are tomatoes, peppers, and members of the onion family: garlic, leeks, and onions.

Artichokes—Clean and trim the hard leaves and sharp tips, then quarter the chokes. Bring to a boil water infused with lemon juice and cook the pieces until nearly done. Place the pieces on a tray and proceed with drying after the cooking water has drained from the pieces.

Asparagus—Dry only the most tender spears (usually the thinnest in diameter), using just the first 3 inches. Blanch until fork tender, but not mushy, about 5 minutes.

Beans, Dry—Let dry in a warm, dark, dry area, such as an attic, in the pods. Shell when the pods are paper dry and store in moisture-free containers.

Beans, Shell—Shell firm, ripe beans (not overmature) and blanch for 5–10 minutes.

Beans, Snap—Cut ripe or nearly ripe beans into bite-sized sections and then blanch for about 3 minutes.

Beans, Soybeans—Drying is not preferred; soybeans need to be fully cooked before eating because they contain an enzyme

A bumblebee harvests the nectar from an artichoke flower at Fat Rooster Farm. The artichoke was overlooked by the author and went to seed. "By then, it's October and I'm sick of harvesting," she said.

that renders them virtually indigestible if not cooked by tradition-al methods.

Beets—Depending upon the size, you will need to blanch beets for 45 minutes or until thoroughly cooked. Remove tops and roots prior to blanching, and after they're cooked, cut them into small pieces or thinly slice them.

Beet Greens—Remove the ribs from the leaves and dry them, unwashed, until they crumble easily. A dehydrator is really the preferred method to dry these and other leaves such as spinach or chard.

Broccoli—Cut florets off the thick, woody stalks (save these by freezing for stock). Cut into strips and blanch between 3 and 5 minutes.

BELOW: **A partial bushel of cleaned garlic is ready for customers at Tunbridge Hill Farm. Farmer Wendy Palthey said the heavy clay soil of central Vermont is ideal for the growth of root vegetables like garlic, onions and leeks due to the soil's ability to retain moisture.**

Brussels Sprouts—Cut in half through the center, and blanch pieces about 10 minutes.

Cabbage—Cut into long, thin strips and blanch for 2-5 minutes.

Cardoon—Remove tough strings and membranes from stalks and cut into small, thin pieces. Blanch for 5 minutes and remove all cooking water before proceeding.

Carrots—Scrub clean, but don't peel; slice into rounds or quarter. Blanch for 5 minutes. These are very good dried, as their sugar content intensifies. They're a great substitute for salty, fatty potato chips from a bag, and they can be rehydrated for soups.

Cauliflower—Prepare as for broccoli.

Celery—Choose slender, tender stalks. Remove the tough, outer leaves (you can dry the inner leaves). Blanch for 2 minutes after cutting into small pieces.

Corn—Remove the husks and silks and blanch the ears for 4-5 minutes. Cool and shave the kernels from the cobs. Dried corn can be stored more easily than frozen or canned corn, and it makes a great trail mix or can be ground into flour.

Eggplant—Peel the outer skin and slice thinly, about ½ inch thick. Blanch for 2½-4 minutes. An alternative method is to cover the sliced rounds with kosher or other coarse salt for several hours to remove moisture, then proceed with drying.

Garlic—Dried garlic is extremely versatile; its flavor mellows a bit so that the sharp sting won't crowd other flavors when used in soups, salad dressings, or marinades. There isn't any need to blanch garlic; simply peel and thinly slice before drying.

Horseradish—Once this root is grated or sliced or heated, the sharp, hot flavor will quickly deteriorate. For this reason, it is usually preserved by combining it with salt and vinegar and kept refrigerated. To dry, don't blanch, simply trim the tops and grate or slice.

Kale—Prepare as for beet greens.

Leeks—Peel off tough outer leaves. Trim to where greens begin to get pale and slice thinly. Don't blanch.

Mushrooms—Gently rinse and slice thin. The fastest method to dry mushrooms is in the dehydrator, requiring only about an

Dried beans in the pantry at Donna Kausen's home. Kausen and her two friends grow more than a half-dozen varieties for soups and other dishes.

ABOVE: As a symbolic guard of the house, a carving that was given as a gift is on an exterior pantry wall where bottles are cemented into the structure at Donna Kausen's home.

hour, though they can also be dried on a cookie sheet in a warm oven or strung to dry in a cool, dry room. No brining or added seasonings are necessary; store dried mushrooms in airtight containers to keep out moisture.

Okra—Use only small, tender fruits. Cut off tips and tops and slice, then blanch for about 5 minutes.

Onions and Shallots—Peel outer dry papers, then finely chop (don't blanch).

Parsnips—Prepare as for carrots.

Peas, Shell—Prepare as for dry beans.

Peppers, Hot—Slice them for a dehydrator (don't blanch) or put a string through them to hang. I like to pick hot peppers when they have turned their mature color; they keep better. You can pull up whole plants with the peppers still attached if frost is threatening. After they've dried completely, store them in airtight containers to avoid dust and household debris. Use dried peppers in Nahm Jeem Gratiem and Sriracha sauces (see page 32 and page 99).

Peppers, Sweet—Pick full, ripe peppers (don't need to blanch); clean, devein and seed, then slice thinly. Dried sweet peppers, particularly paprika varieties, are good ground into fine powder using a food processor or herb mill.

Potatoes—Not to be confused with potato chips, dried potatoes are brittle and bland tasting, but they're great additions to soups and stews. Peeling the potatoes is not necessary; clean and slice ½ inch thick. Blanch for about 5 minutes, then plunge them in cold water with lemon juice (about 1 cup to 1 gallon of water).

Pumpkin and Winter Squash—Seed and remove stringy insides. Slice into wedges about 1–2 inches thick and peel.

Rhubarb—Cut stalks into thin strips and blanch 2–5 minutes before drying. Never eat or dry the leaves; they're poisonous.

Spinach—Prepare as for beet greens.

How Best to Store Your Mushrooms

While some varieties of mushrooms can be successfully canned or frozen, the best way for long-term storage of mushrooms is through drying. If you're foraging for wild mushrooms, be sure that you've properly identified the specimens that you collect; there are many fatally poisonous varieties out there, and just as many that could cause severe gastrointestinal problems.

Select only firm, young specimens (leave the older wild varieties to spread their spores for next year's harvest). They can be gently rinsed, and in the case of larger varieties like morels or boletes, they can be halved. Watch particularly for uninvited guests like slugs or snails, small earthworms, or crickets.

To reconstitute, soak the mushrooms in hot water until softened, about 20 minutes. Don't throw out the liquid used to cook them! About 3 ounces of dried mushrooms will be the equivalent of 1 pound fresh.

Summer Squash and Zucchini—Slice into ¼–½- inch rounds without peeling. Blanch for 2–3 minutes. If you're using a dehydrator and eating them as chips, there's no need to blanch (they won't stay crisp for much longer than a week, though, so blanch them if you're going to store them longer). The squash can also be sliced into thin strips

Sweet Potatoes—No need to peel unless bruised; wash then grate or dice for longer storage. Blanch for 3–8 minutes. If eating as chips within a week or two, don't blanch, simply slice into ¼-½-inch rounds and use a dehydrator.

Swiss Chard—Prepare as for beet greens.

Tomatoes—Besides hot peppers, tomatoes are one of the most common of the dried vegetables (okay, technically it's a fruit). Paste tomatoes are the meatiest, so they dry the fastest if you're air-drying them, but if you have a dehydrator, you can dry any variety quickly

BELOW: **Peppers are gathered for distribution to Community Supported Agriculture (CSA) members at Sunrise Farm in Hartford, Vermont.**

and efficiently. I like to slice mine thin, ¼-½ inch thick, and dry them in a dehydrator. If you intend to use them as sun-dried tomatoes marinated in olive oil, they should be dried slowly over at least a 24-hour period. They will be leathery, not brittle or hard. To make tomatoes that are flaky and dry, use a dehydrator set according to the manufacturer's instructions. Dried tomatoes can be chopped into flakes using a food processor and added on top of salads, meats, and pastas. Store in airtight containers.

Drying Miscellaneous Foods

Yeast or Sourdough Starter—Prior to 1876, when Charles L. Fleischmann made his debut at the Centennial Exposition in Philadelphia, most leavened baking was done with *starters* or *sponges* cultivated from naturally occurring yeasts and bacteria in the air. Early colonial cookbooks are filled with starter recipes, ranging in ingredients from hops and potatoes to whole-wheat flour and sugar. Several starters are famous; the San Francisco sourdough is probably the most well known, making claims such as the French do, that even if the sparkling wine is made from Champagne grapes, if it's not made in Champagne, France, it's only sparkling wine.

Starters can be frozen or dried successfully, which might make them more manageable for the modern-day kitchen—they're labor intensive unless you bake consistently. They need to be fed regularly and babied a bit fresh.

To dry the starter, shape it into small balls (they'll probably flatten out during the drying), and dry them in the sun, in the dehydrator at less than 120 degrees Fahrenheit, or next to the woodstove. After the balls have dried, finely grind them in a spice grinder and store in a sealed glass jar or in the refrigerator in a plastic bag.

To refresh the starter, take about a tablespoon-full out and add water and a little food (flour and honey or mashed potato). Put in a glass container, loosely covered, in a warm area. Within 2 or 3 days, it will pick up and become starter again.

The door to the root cellar at Earthwise Farm and Forest was part of the foundation's design when the house was built; farmer Carl Russell said he hasn't finished the header storage compartment over the door. He joked that now it's holding all of the farm's profits.

Canned items put away on the shelves in the basement of Fat Rooster Farm include pickled beets and peaches (top shelf), more beets, black currant jam, rhubarb jam and maple syrup (middle shelf), and stewed tomatoes for sauce (bottom shelf). At the bottom are carboys for fermenting wine and beer.

(4)

Canning Fruits
and Vegetables for
Storage

BEFORE MODERN REFRIGERATION and freezing were household norms, the pantry was more important than having indoor plumbing. In the farmhouse we dwell in, built in 1872, the pantry resided in fully one quarter of the downstairs, tucked into the northeastern corner, where cool, dry conditions prevailed without as much fluctuation as other parts of the house, making it the perfect spot for canned goods.

Canning in History

Before the early 1800s, foods were preserved by storing them in containers and sealing them with oil or fat so that no airspace remained. In medieval Europe, foods were sealed when fat was poured

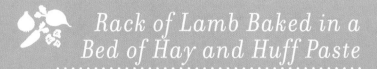

Rack of Lamb Baked in a Bed of Hay and Huff Paste

Cooking history is fascinating to me. Before the convenience of the fridge, creative cooks used a myriad of methods to preserve their foods. Not only does this inedible outer paste keep the meat from spoiling because it has been sealed tightly, it keeps the juices inside instead of burning on the bottom of the pan. Don't use hay from a pet store; get it from a farmer, and if you can't, substitute the shuckings from corn on the cob.

Rack of lamb
Hay or corn shucks to wrap the meat
2 sprigs of rosemary
6 cloves garlic, peeled and slivered
3 tablespoons fish sauce or good quality anchovies
1 egg, beaten
Kosher salt and freshly ground black pepper
4 cups all-purpose unbleached flour
½ cup coarsely chopped beef suet
3 teaspoons kosher salt
2 teaspoons crushed, dried mint leaves

into the hole in the crust of a pie until it was filled. The food beneath the crust and fat was preserved, but the crust itself, a mixture of coarsely ground rye flour, water, and a little suet, called huff paste or coffin (see recipe above), was only eaten by the house servants, who devoured it for the meat and vegetable juices it had soaked up during cooking. The waste of the crust was avoided later, when it was found that earthenware pots could be used to cook the food, and a layer of hot fat was poured on top to seal its contents. Spoilage still occurred occasionally if the food cooled before the fat sealed the surface, or the fat cracked and allowed air to enter the pot.

It wasn't until Nicolas Appert, now referred to as the Father of Canning, began experimenting in the early 1800s with preserving

Add enough water to the flour, suet, and, spice mixture to make the dough rollable. Knead the dough by hand for 15 minutes, then let it set overnight in the refrigerator. Soak the hay or corn shucks in cold water overnight.

Roll out the dough to approximately ¼ inch thick over a baking tray. Squeeze out the hay or corn shucks and place them on the pastry in a thin layer, leaving about a 2-inch perimeter at the edges. Sprinkle the herbs on the hay or corn shucks. Brush the edges of the dough with olive oil or water.

Preheat the oven to 375 degrees Fahrenheit. Make small slits in the rack of lamb and insert the garlic slivers. Douse the rack with the fish sauce, or place the anchovies between the rib bones. Season the rack with salt and pepper, and lay the rosemary sprigs on top.

Put the rack on the dough and wrap the dough over the meat, leaving one or two rib bones protruding. Pinch the dough tightly shut, seal it with the beaten egg, and let it rest, refrigerated, for an hour.

Bake the pastry until the internal temperature reaches 145 degrees Fahrenheit for medium rare meat. Let it rest for ½ hour before cracking open the huff paste. The crust is inedible, though the family chicken flock will likely enjoy it. This goes well with roasted root vegetables doused in olive oil and sprinkled with marjoram or oregano.

foods in glass containers, sealed with heavy cork and heated at high temperatures, that progress was made preserving foods through canning. Encouraged by Napoleon Bonaparte's wager of 12,000 francs to the man who could come up with a practical method of food preservation for his armies, Appert worked for more than 10 years experimenting with preserving food. In 1810, he published his book, *L' Art de conserver les substances animale et vegetales*, and collected the prize. Appertization, should not be confused with Pasteurization, however. It wouldn't be until several decades later that Louis Pasteur would discover and describe scientific developments that would drastically change food preservation techniques. He realized that sealing out the microorganisms within the containers and exposing them to heat that also deactivated enzymes within the food would preserve it for several years. Pasteurization occurs at a much lower temperature than Appertization, and it also better preserves the nutrients and integrity of the foods being processed.

Concurrently, Peter Durand was working on a means by which to preserve food in metal cans. He sold the patent to his idea in 1811 to Bryon Donkin, who with John Hall began to manufacture cans out of heavy iron. Each can needed to be hand-made, and the food inside was processed for up to 6 hours. The cans were so thick and cumbersome that directions for opening them were printed on the can and read *Cut round the top near the edge with knives, a chisel and hammer or even rocks* (Ezra Warner didn't invent the can opener until 1858). Metal cans were used predominantly on the battlefield, not in the home kitchen; it was considered a status symbol and a luxury to serve canned food in the household.

When World War I broke out, the American government embarked upon a campaign to galvanize its citizenry to "Back up the cannon with the canner." The goal was to inspire the average housewife to use canning as a way to preserve the harvest rather than rely upon commercially prepared canned products in the home. That way, the military could utilize the commercial canned products on the battlefield.

By the late nineteenth century, metal canneries were much more common; Nestlé, Heinz, and others had factories throughout the United States and were catering to urban markets. Cans were easier to manufacture than glass jars, easier to transport, and much less fragile than their glass counterparts. However, for the home canner, there were several disadvantages: They required specialized, expensive equipment, they were expensive at the home canner's economy of scale, and they weren't reusable.

In 1858, John Mason of New York City invented a machine that could cut threads into the glass jar whereupon a metal ring and lid could be secured. The ease and affordability led home canners to turn to these, as well as products from Ball and Lightning (invented by Henry William Putnam of Bennington, Vermont, and featuring

Canned tomatoes from the previous season and empty jars are stored in the cellar at Fat Rooster Farm. The stewed tomatoes were to be used for spaghetti sauce. The cool, wet summer of 2009 brought a tomato blight that killed 600 plants on the farm—no tomatoes remained for canning in the fall.

a glass lid with a wire bail) to do their home preserving. Today, it is still possible to purchase canning equipment easily and affordably. For many without the option of refrigeration, it is still the method of choice by which to put food up.

Storing Canned Goods

If you're using canning jars with metal lids and rings, remove the rings before storing them. Wipe the jars down with a damp cloth, and check to see if there is food between where the ring was seated and the metal lid. Use these jars immediately; store them in the refrigerator.

For properly sealed jars, be sure to label the contents with the name of the food and the date it was processed. Using the jars within two years, and ideally one year, will prevent chemical changes in taste and appearance from occurring. Food stored longer than two years in jars will have reduced nutritional value, as well. Store them in a clean, cool (between 40 and 65 degrees Fahrenheit is ideal), dry spot. Freezing temperatures can cause the jars to crack and break; temperatures above 90 degrees Fahrenheit can cause spoilage (woodstoves, radiators, and furnace vents as well as direct sunlight can also spoil the jar's contents).

Most root cellars are too humid for long-term storage of canned goods because the moisture in the air—the same moisture that keeps the root vegetables stored beautifully—will tend to rust the metal rings and lids of the canned goods after about a year. The rust will create small openings into the canning jar and then food spoilage can occur. This can set you up for some seriously bad food-borne illnesses, including the most deadly (and most rare), botulism. That said, you can modify your root cellar by dividing off a section that may be drier due to the geography of the ground, and store your canned goods there. Or make certain to check them for any discoloration, leaks, or off smells, and discard anything that hasn't been eaten in a year.

To convert a space in a newer dwelling to a pantry, choose a cool, dry space such as a corner in the attic or an unheated, upstairs bedroom. Such spots are a good location for canned goods, onions,

LEFT: The author uses old and new types of canning jars for pickling.

shallots, garlic, winter squash, and pumpkins, all of which prefer temperatures in the 50s and 60s Fahrenheit and humidity less than 60 percent.

How Canning Preserves Food

This chapter will deal with techniques that use glass canning jars, not metal cans. If you are thinking of starting a value-added business and are considering canning your product in a tin can (actually 100 percent recycled steel these days), you should also take into consideration marketing. Often, glass jars are thought of as classier; they are more colorful, the consumer can see the product within the jar, and the jar can be recycled *or* reused. In the long run, you may come out ahead using glass.

Canned food consists of raw or cooked ingredients that are packed into jars and filled with enough liquid that most of the airspace inside the jar is displaced. Lids and rings (either glass lids fitted with rubber rings and wire bails or metal lids with metal rings) are then fastened tightly on the jars. The jars are then heated to kill microorganisms and deactivate enzymes inside; at the same time, air is forced out of the jars with the pressure of heating. Depending on the acidity of the food being canned, either a hot-water bath or a pressure canner is used to process the food for a specific amount of time. The jars are then removed from the processor and left to sit undisturbed for at least 12 hours, while the seals are set and the food cools inside the jars.

Canning Equipment

Obviously you'll need the jars themselves, ideally modern jars that seal with metal lids and rings. Try to avoid ceramic-lined lids made with zinc alloy, not the highly oxidizable tin of today, and jars with glass tops and rubber rings. These have been deemed too risky to use, partly because the signs of proper sealing were too subtle to detect for the modern canner, and partly because it was thought that the wire bails on the lids kept the tops from lifting off the jar, even if harmful bacteria were producing toxic gases within (keep in mind that most of Europe still uses these glass-lidded, rubber rimmed jars, without fail).

Nowadays, home canners rely on the conveniently indented metal lid of a properly sealed jar and the distinct pop upon opening it to ensure that the contents inside have been properly sealed and preserved. This is far easier to navigate than the subtle whoosh of the rubber seal upon opening that the glass-lidded canning jars of old provided. Curiously enough, the rubbers are still available for sale in many areas of the United States where they have been condemned, as are the glass jars. Canning jars with metal lids and rings are by far easier to deal with: They're easier to know when properly sealed, and easier to open when you want to use the food. They're also more reliable when it comes to sealing. However, jars with glass lids have been used for decades, so simply be aware of the

ABOVE: **Tools used for canning and drying herbs and vegetables are shown.**

dangers if you choose to use them. See Resources for modern-day purveyors of canning supplies.

Depending upon the food you are canning, you'll need either a boiling water canner or a pressure canner (either a weighted gauge or a dial gauge). An average sized canner can hold seven quart jars or up to nine pint jars. Pressure canners (not to be confused with pressure cookers, which are generally smaller in diameter) will give the home processor many more options for processing mixed vegetables and other low-acid concoctions. If you're new at canning, you might want to start out with a boiling water canner and high acid foods and then advance to the pressure canner when you feel confident.

Use a boiling water canner if you're processing foods like most fruits, pickled and fermented vegetables, and tomatoes. You can use any large pot to can these products; mine is an old-fashioned water bath pot, the same one that I use to boil lobsters in, with a nonreactive surface. You can use your pressure canner to preserve

Empty and full canning jars are stored in the farmhouse cellar at Fat Rooster Farm.

Should I use my commercial glass jars from foods like spaghetti sauce and salsa to can my own products?

I wouldn't recommend it. The walls of commercially made, one-time-use jars are thinner than the jars made for home canning. The rims can be chipped when the jar is opened and being used, and they can form hairline cracks that cause them to burst when processed a second time in a hot-water bath, and I would never risk canning with them in a pressure canner. Having said this, it's perfectly fine to use such jars to store dried foods, such as mushrooms, lentils, flour, and herbs.

food using a hot-water bath; just remove the petcock if you have a dial gauge, or the weight, so that any pressure created by boiling can be released through the steam. Make sure that the water inside covers the tops of the jars at least 2 inches above, and that it doesn't reach to the top of the pot (there won't be any room for the water to boil, and it'll create a mess on the stove). There should be a rack in the pot so jars are held off the bottom (sometimes I cheat and put a dish towel under the jars—if you do this, you run the risk of the jars being agitated enough that they will crack).

Other Tools You'll Need

Along with jars and a canner, there are several other useful canning utensils. A **canning funnel** fits snugly just inside the jar's rim so that food can be safely packed inside the jars without soiling the rim. A **lid wand** is a long, usually plastic, wand that has a magnet attached to the end to grab onto jar lids and rings from their boiling water sterilizing bath. A long thin **spatula** is handy to run down the inside of the packed jars, eliminating air pockets (a butter knife works, too). A **jar lifter** is used to take the jars out of the canner after being sterilized; it's handy to put them back in after they're filled and to take them out after processing. If you're hot-packing foods, like spaghetti sauce, you'll need a large, heavy-bottomed, nonreactive pot in which to cook the food prior to canning.

Hot-water Bath versus Pressure Canning

Foods can be preserved through canning by processing in a hot-water bath (a canning pot that contains boiling water where jars are "bathed" over several minutes until sealed) or in a pressure canner that can achieve much higher temperatures than a normal pot of water can, thereby destroying potentially lethal bacteria such as *Clostridium botulinum*. The acidity (or pH) of the food canned will determine which method is suitable for your preserving.

On the pH scale, the higher the number, the less acid is found in the food. All foods with a pH higher than 4.4 are considered a low-acid food and should not be processed in a water bath due to the potential of bacteria to thrive and grow within the sealed jar (the high acid prevents this growth). Tomatoes have always been controversial, because some varieties, especially some of the heirlooms, have a pH just a hair higher. Most canning manuals recommended

When Not to Use Canning to Preserve Your Food

Some foods don't do well at all being canned. They turn mushy or have an off color after becoming subjected to the heat. In general, I choose canning for fruits other than berries (unless you're making them into jellies and jams), tomato products, and pickled foods, from eggs to dilly beans. Any of the cole crops or brassicas (broccoli, brussels sprouts, cabbage, cauliflower, and kohlrabi) unless they're being pickled, are bland when canned and become grayish; eggplant isn't good canned unless you're making it into caponata relish; celery, celeriac, turnips, parsnips, and rutabagas fare much better in the root cellar whole and used fresh. Patty pan squash, summer squash, and zucchini do better frozen unless you're making them into pickles.

adding a bit of lemon juice or cider vinegar to each jar to be safe that the pH is low enough. Fruits and pickled vegetables are fine with a water bath but all other vegetables, or mixed vegetables that are not pickled, should be processed using a pressure canner.

Pressure canning is the only method recommended for canning low-acid fruits, vegetables, fish, and meat. Each pressure canner will come with a set of instructions on proper use of the canner; some canners have weighted gauges to regulate the pressure inside, while others have a dial gauge. Dial gauges generally gauge pressure more accurately.

At sea level, the temperature of the canned food inside should be maintained at 240 degrees Fahrenheit, for the specific amount of time as indicated for each fruit or vegetable. If you're at higher altitudes, say above 1,000 feet, you'll need to adjust your canning times accordingly; the temperature of boiled water or steam is actually lower at these altitudes. This means that you'll have to increase the time you boil in the hot-water-bath canner or increase the pressure in steam pressure canners. See chart on page 133.

The canning rack should be placed in the bottom of the canner, and then it should be filled with about 3 inches of hot water. Some canners can be stacked double-decker, so that a rack is placed in between and another row of jars is placed on top. Adding vinegar or lemon juice to the water can prevent the canner from turning color; 2 or 3 tablespoons is sufficient.

Put the lid on the canner and lock it securely in place. Keep the vent open, put the heat to its highest setting on your stove, and let it stand until steam begins to erupt from the vent. Let it vent for at least 10 minutes at sea level, or longer at higher altitudes (see below). Then close the vent and bring the canner up to the correct pressure.

If you have a weighted-style pressure canner, follow your canner's instructions to use the proper weights and learn how to determine when the correct pressure has been reached inside the canner.

Dial gauge canners should be kept on high heat until they've almost reached the correct pressure (about 3 pounds below the desired pressure). Then switch the heat to medium until the correct pressure is reached.

When you've reached your desired pressure, begin timing the processing, keeping a close eye on either the weight, which should be rocking back and forth, or the dial gauge. Don't allow pressures to fall below the correct pressure, or you'll have to start again by waiting until the correct pressure and then timing for the full amount required by the food. Pressures shouldn't exceed 15 pounds, and try to keep the pressure even so that the food isn't sucked out of the jars, ruining the seals.

After the food has been processed for the proper amount of time, turn the heat off, take the canner off the burner, and let it sit until the pressure registers 0. This may take a while, perhaps an

BELOW: The canned and harvested vegetables stored in Donna Kausen's root cellar lasts the season for her and the two friends she shares her property with. Some years, they even have enough to either give away or feed to their animals when it's time to put away more of the harvest.

Canning in Higher Attitudes

Adjusting Canning Time and Pressure for Higher Altitudes

Most canning recipes are formulated for processing at or close to sea level. Because the temperature of boiling water or steam is actually lower the higher you are above sea level, you must increase the time the food is processed in the hot-water bath or increase the steam in the pressure canner to effectively can food safely.

HOT-WATER BATH PROCESSING

0–1,000 feet, follow recipe
1,001–3,000 feet, add 5 minutes
3,001–6,000 feet, add 10 minutes
6,001–8,000 feet, add 15 minutes
8,001–10,000 feet, add 20 minutes

PRESSURE CANNER PROCESSING

Weighted Gauge

0–1,000 feet, use 10-pound weight
1,001–10,000 feet, use 15-pound weight

Dial Gauge

0–2,000 feet, use standard pressure of 11 pounds
2,001–4,000 feet, 12 pounds
4,001–6,000 feet, 13 pounds
6,001–8,000 feet, 14 pounds
8,001–10,000 feet, 15 pounds

hour. Then you can take the weight off or open the vent, depending on the type of canner. Wait another few minutes before removing the canner lid. Carefully remove the jars.

Preparing the Jars for Packing

Glass jars come in a wide variety of widths at their openings. Most standard jars made by Mason and Ball are referred to as either regular-mouth or widemouthed. Weck, a company out of Illinois, United States, offers unique canning jars without a bail lid. The width of the jar's opening varies, according to jar type.

Emma Hansen tastes a strawberry while picking them with her family at 4 Corners Farm.

To prepare the jars for packing, wash them and the rings and lids with hot, soapy water (either by hand or in a dishwasher) to remove any dirt, dust, or oily film if they've been used before (it's not a good idea to reuse metal lids, though the rings can be used again if they aren't rusty).

If the food is to be processed for less than 10 minutes, the jars need to be sterilized. This includes all pickled products, jams, or jellies. To sterilize the jars, put them in the canner in hot water (never put a cold, even room-temperature, jar in boiling water—it could explode) to an inch above their tops. Bring the water to a boil and boil for 10 minutes, unless you are at altitudes of more than 10,000 feet. Then, increase the boiling time 1 minute for every additional 1,000 feet of elevation.

Remove the jars from the boiling liquid and drain (save the water to process the jars once they've been filled).

If you're processing pickled or fermented foods or tomatoes and fruits for longer than 10 minutes in a boiling hot-water bath, there is no need to sterilize the jars, nor is there a need to sterilize jars that are to be processed in a pressure canner. Just make sure they are clean.

Headspace

A certain amount of room must be left between the top of the contents and the lid so the food can heat up, swell, and move around. Without this headspace, the contents can run up and out of the jar, spoiling the seal. However, too much room results in excess air that will prevent a good vacuum during processing, necessary for a proper seal. Generally, about $\frac{1}{2}$ to 1 inch of headspace is needed; refer to individual foods for the proper amount required.

Packing the Jars: Hot-Pack versus Raw-Pack

Hot-pack—processing is when hot, sterilized jars are filled with precooked food and boiling liquid. Advantages to hot packing are that air is forced from the food being processed by cooking, thereby improving quality (air provides a medium for bacteria to grow and multiply, resulting in spoilage); you can fit more food into the jar;

the colors of the foods are better preserved, and more nutrients are retained, because there's less potential for oxidation to occur (there's less air inside). Hot packing is strongly recommended for all water-bath-processed foods. I don't like to hot-pack any foods being pickled—it takes the crunch out of them. I do heat the pickling liquid to simmering, though.

Raw-packed—foods are packed into hot, sterilized jars and filled with boiling liquid.

For both methods, have the liquid ready to fill in after the food has been packed into the sterilized jars. Don't overfill the jars—allow them to fit in the jar snugly, but give the food some wiggle room. Sliced fruit or vegetables can be layered for the best fit; I like to load dilly beans and asparagus by tipping the jar on its side so the spears or beans are arranged vertically.

Checking the Jar Seals

After the canned food has cooled, gently press down on metal lids. They should be curved slightly toward the contents and not pop or give at all when pressed. If they do, simply refrigerate these jars and eat them within a week's time. Glass-lidded jars should stay

Safe Canning Practices

Canning will go without a hitch if some general rules are followed.

- Don't use a hot-water bath to can low-acid foods. Although small, the risk of botulism poisoning is there. Using a pressure canner will eliminate this concern.
- Don't reuse lids or rubbers. The seal may have been compromised during the last use.
- If a jar is bulging at the lid or leaking food, throw it out. Don't test it! If you notice food leaking immediately after canning, refrigerate the cans and use immediately.
- When you open a properly sealed jar, it should be a little difficult. You'll hear a sucking whoosh when the lid or rubber is loosened.
- With borderline foods, like low-acid tomatoes (many of the heirlooms and overly ripe tomatoes tend to be less acidic), use an additive like cider vinegar or lemon juice.

Table 3: **How Much Fresh Food Will Yield When Canned**

Raw Food	Pounds of Raw Food Needed Per Quart
Apples	3
Apricots	2.5
Beans, dry	0.75
Beans, snap	2
Beets	3
Berries	3
Carrots	3
Cherries	2.5
Corn	6 to 7 ears, kernels cut off cob
Okra	2
Peaches	2 to 3
Pears	2 to 3
Peas	4.5
Peppers, Hot or Sweet	I pound per pint
Plums	2.5
Potatoes	5
Pumpkins and Winter Squash	2.5
Summer Squash	2.5
Spinach	3 (use this amount for other greens)
Sweet Potatoes	3
Tomatoes	3.5 pounds whole or halved; 6 pounds tomato sauce

snugly against the rubber rings and have nothing leaking from them. Label all properly sealed jars with the contents and the date processed.

When you open a jar of canned food, don't throw out the juice inside. During the canning process, many of the vitamins and minerals are leached out of the solids into the surrounding liquid. Use sweet liquids from fruits diluted with water to make light juice; use heavier syrups over ice cream or cereal in place of milk. It's also best to bring home-canned foods to a boil before eating them, just to make sure that undetected bacteria are destroyed.

How to Can Fruit

Fruit is probably one of the easiest foods that home canners can prepare. Fruits are typically high in acid, so spoilage due to overgrowth of low-acid loving bacteria can't occur, and the canned product is generally high quality in both taste and appearance.

Apples—I don't bother turning my apples into anything other than applesauce or cider, but they're easily canned as slices. Use firm, crisp apples, mixing sweet with tart for maximum flavor. Wash them, peel and core, and slice them about ¼ inch thick into an anti-browning solution (see page 140). Boil gently in boiling water for 5 minutes, and pack into hot jars. Top off with the cooking liquid, leaving ½ inch headspace. Process in a hot-water bath for 20 minutes.

Apricots, Nectarines, and Peaches—I prefer to peel off the fuzzy skins (you don't have to peel nectarines), but you can skip this step. As you would tomatoes, drop the fruit in boiling water for 30 seconds to 1 minute. Remove them with a basket or slotted spoon, plunge into cold water, then slip off the skins. Pit the fruit and remove the dark red lining where the pit was (it gets dark when canned); coarsely chop the fruit. Cut into wedges and dip in an anti-browning solution.

The fruit can be packed in a light syrup (see page 141), but I prefer to can them in their juice or with water, so that I can optionally add the sugar as desired later when they're used.

Bring the drained fruit to a boil in a large saucepan or pot. Ladle the fruit into hot jars using a basket or slotted spoon, and top off with the liquid, leaving ½-inch headspace. Process in a hot-water bath for 20 minutes.

Berries—Many home canners claim that the delicate flavor of berries like blackberries, blueberries, elderberries, gooseberries, mulberries, and raspberries is sacrificed when they're canned. Strawberries really don't can well, and are far better frozen. Perhaps if freezing is not an option, canning their juice or turning them into preserves results in a better final product.

Hot packing is the preferred method of canning whole berries, and to every quart of washed, stemmed, and drained fruit, ¼ cup of honey or ½ cup of sugar should be added.

Put the fruit and syrup in a pan, cover, and bring to a boil. Stir a few times, and then ladle hot berries into hot jars, leaving ½ inch headspace. If there's not enough juice to cover the berries, cover with the cooking liquid or boiling water. Process for 15 minutes.

Cherries—Canning is one of the best ways to enjoy cherries out of season. Sour and pie cherries can be pitted before they're frozen, but their color won't keep as well. Unpitted must be pricked with a clean pin or they might burst. Stem the fruits and wash them. If pitting, put them in an antibrowning solution of 1 teaspoon of lemon juice per gallon of water. Drain before processing.

Add ¼ cup honey or ½ cup sugar with each quart of fruit. If unpitted, place in a pan and add a little liquid (pitted cherries will usually make their own juice in the pan) to prevent scorching. Cover and bring to a boil. Using a basket or slotted spoon, pack hot fruit into hot jars and add hot cooking liquid, leaving ½ inch headspace. Process for 15 minutes for pints, 20 minutes for quarts.

Cranberries—Canned cranberries are softer than if they're frozen, but they make excellent sauces for meats and toppings for

 Keeping the Brown Out

Many light-colored fruits will discolor after they've been bruised (or cut up), much like potatoes will. To prevent this, you can add an anti-browning solution or acidifier. There are commercial anti-browning mixes available, but it's easy to use a 500-mg vitamin C (ascorbic acid) tablet that has been crushed in 1 gallon of water or 1 teaspoon of lemon juice per gallon of water. Drain the fruit well before canning it, but don't rinse.

··········· Using Syrups for ··········· Canning Fruits

Many home canners prefer to top off their fruits in a sugar solution or syrup. Since we tend to use more maple syrup and honey on the farm, I prefer to add the sugar when I'm using what I've canned. Be sure to make the syrup before you've started to can the fruit or your timing will be off. Following are recipes for 10, 20, and 30 percent sugar syrups.

10 Percent Syrup: Add ¹/₃ heaping cup sugar and enough water to make I quart of liquid.

20 Percent Syrup: Add ¾ heaping cup sugar and enough water to make I quart of liquid.

30 Percent Syrup: Add 1¼ cups sugar and enough water to make I quart of liquid.

sweets. The tart fruits should be packed in sugar syrup. Combine 1 ²/₃ cup sugar and enough water to make a quart. Stir well to dissolve the sugar; allow ¾ to 1 cup of syrup for each pint of fruit.

Figs—I think figs are better dried, but they have a nice texture when canned (they lose flavor over time, so use them within a year of canning). Choose only firm, ripe (not overripe), uncracked fruits. Wash the fruits and then drain—don't bruise the fruit by stemming or peeling it. Figs are borderline low-acid, so an acidifier must be added. Cover the fruits with water and boil for 2 minutes. Drain, then boil gently for 5 minutes in water, juice, or syrup made from ¾ cup sugar and enough water to make a quart (allow ¾ cup to 1 cup of finished syrup per pint of fruit). Transfer the hot fruit to hot jars and add 1 tablespoon of lemon juice to each pint jar. Top off with the cooking liquid, leaving ½ inch headspace. Process for 45 minutes for pints, 50 minutes for quarts.

Canned peaches are
stored in the cellar
of the house at Fat
Rooster Farm.

Grapefruits, Oranges, and Tangerines—Don't can oranges by themselves; they just don't taste as good as when they're combined with other citrus. These fruits are best raw-packed. Syrup, juice, or water can be used to top off jars. For syrup, use ¾ cup sugar dissolved in enough water to make one quart; allow ¾ to 1 cup of syrup to each pint of fruit.

Wash, peel, and remove the white pith covering the fruit. Separate into segments, removing the seeds if possible.

Bring the liquid of choice to boil. Pack the raw fruit into hot jars and add the hot liquid, leaving ½ inch headspace. Process for 10 minutes.

Grapes—Grapes that are available in stores are typically seedless and thick skinned. "Slip skinned" grapes are cultivars more likely to be grown by home gardeners or found wild. These typically have skins that easily pull away in one piece or "slip" from the fruits. They have seeds, but can be successfully canned. To avoid the skins and seeds, prepare wild-type grapes as juice or preserves. Otherwise, pick grapes slightly underripe, wash, and stem. Bring syrup, juice, or water to a boil in one pot and a gallon of water in another pot. If using syrup, dissolve a heaping ¾ cup of sugar in enough water to make one quart. Allow ¾ to 1 cup syrup to each pint of fruit.

In the water pot, dip the grapes in the boiling water using a colander or blanching basket. Drain and then pack the grapes in hot jars, and pour the hot packing liquid over the grapes, leaving ½ inch headspace. Process for 10 minutes.

Papayas—These fruits are borderline low-acid, so adding 1 tablespoon of lemon juice to each pint jar is recommended.

Wash and peel fruits. Cut them in half, remove the black seeds, and cube. Prepare a syrup by combining 1¼ cups sugar with enough water to make 1 quart of syrup; allow ¾ to 1 cup of syrup to each pint of fruit. Bring the syrup and the fruit to a simmer and cook for about 3 minutes.

Remove the fruit with a basket or slotted spoon and pack into hot jars. Top off with the hot syrup, leaving ½ inch headspace. Process for 15 minutes.

Peaches—See Apricots, Nectarines, and Peaches.

Pears—Wash and peel fruits, then halve or cut into quarters and core. Soak in a solution of 1 tablespoon of lemon juice in 1 gallon of water. Drain and bring to a boil in water or fruit juice (if the fruit is very juicy, just add honey to taste and boil by itself). Cook gently for 5 minutes, then ladle the hot fruit into hot jars with a basket or slotted spoon. Top off with the boiling liquid, leaving ½ inch headspace. Process pints for 20 minutes and quarts for 25 minutes.

Pineapple—Wash the fruit, cut off the stem, and peel off the eyes. Cut into quarters lengthwise, then cut out the core. Slice, chunk, or cut into wedges. Simmer in water or pineapple juice for 10 minutes.

Using a basket or slotted spoon, ladle the fruit into hot jars. Add the hot liquid, leaving ½ inch headspace. Process for 15 minutes for pints, 20 minutes for quarts.

Plums—Wash, then prick whole fruits so they won't burst. Heat the fruit slowly to boiling to release the juices, then let sit 20 minutes in a covered pan. Ladle the hot plums into hot jars, then top with hot liquid (if the fruit is not juicy enough, use boiling hot water or juice), leaving ½ inch headspace. Process for 20 minutes.

Rhubarb—Another fruit that I'd rather freeze for future pies, or turn into chutney, jelly, or jam. Still, if you want to can rhubarb, harvest tender, red stalks (discard leaves—they're poisonous). Wash, cut into pieces, and combine in a pan for every quart of rhubarb ½ to 1 cup of sugar. Let stand for less than 3 hours, stirring occasionally, and then slowly bring the mixture to a boil. Ladle into hot jars, leaving ½ inch headspace, and process in a hot-water bath for 15 minutes.

Strawberries—Not recommended. Freeze or make into jellies and preserves instead.

How to Can Fruit Juices

Making juice at home is simple, rewarding, and a safe project for the beginning canner. The flavor of the fruit is intensified, and it can be sweetened to taste upon drinking. A juicer is the handiest device to juice effectively (be sure to heat the juice before filling the jars), but fruit can also be cut up and simmered, then strained through a colander. You can achieve a less cloudy finished product if you allow the juice to sit for 12–24 hours and pour off the clear juice. The juice is then heated and poured into hot jars (don't boil the juice, or it will diminish its quality). Follow each recipe for specific headspace requirements.

Apple—This is by no means to be confused with apple cider. Until recently, apple cider meant unpasteurized apple juice. The tang and depth of taste of fresh apple cider is nothing like apple juice, nor is it comparable to pasteurized apple cider. Due to food safety regulations, however, unpasteurized apple cider is prohibited for sale in many states.

Use a variety of sweet and tart apples, even crabapples to add bite. I like to throw in a couple of pears just to give the juice zing.

Wash the fruit and cut out the blossom and stem ends. If you have a juice extractor or fruit press, simply juice the fruit. Alternatively, chop the fruit into small pieces and simmer it until it's soft. If you have a Chinois-type fruit strainer, press the fruit through, then let the juice sit for 12–24 hours until it clears. Pour off the clear liquid, bring to a boil, and then pour immediately into sterilized jars, leaving ½ inch headspace. Process in a boiling-water bath for 5 minutes for pints and 10 minutes for quarts.

Apricot, Nectarine, or Peach Nectar—Wash, pit, and chop fruit (no need to peel). Use a juice extractor or cook chopped fruit by adding 1 cup boiling water to each quart of fruit. Gently cook, stirring frequently, just below a simmer. Press the soft fruit through a strainer, add 1 tablespoon of lemon juice per quart of nectar, and add honey to sweeten, if desired. Pour the hot nectar into hot pint jars, leaving ¼-inch headspace, and process in a boiling-water bath for 15 minutes.

Berry, Cherry, and Black Currant—Black currant bushes have an acrid, almost skunky smell about them, and their

RIGHT: Canned crabapples are stored in the pantry at Fat Rooster Farm.

juice is too strong by itself for some people. Try mixing cherries or other berries to the currants to mellow the taste.

Wash, stem, and crush the berries using a potato masher; for cherries, wash, stem, pit, and chop the fruit. Use a juice extractor, or simmer the fruit gently, stirring occasionally. Strain and reheat; sweeten to taste and add lemon juice if desired. Pour into hot jars, leaving ½ inch headspace, and process for 20 minutes.

Cranberry—Discard any soft or discolored fruits, then wash and stem the remainder. Add 4 cups of water for each quart of fruit and bring to a boil. Reduce the heat to simmering, and cook until the berries have all popped. Strain the liquid, then reheat and sweeten with honey to taste. Pour the hot juice into hot jars, leaving ¼-inch headspace, and process in a boiling-water bath for 10 minutes.

Grape—Wash the fruit. Extract the juice with a juicer or cover with 1 cup boiling water for every 7 pounds of grapes. Cook at a gentle simmer, crushing the fruit with a potato masher as it cooks. Strain the juice and let it sit in the refrigerator until the sediment has settled out. Carefully pour off the juice, sweeten to taste, and heat to a simmer. Pour the hot juice into hot, sterilized jars, leaving ¼ inch headspace. Process for 5 minutes. If using white grapes, you will have about half the juice that is produced by purple varieties.

Rhubarb—Use only slender, firm, red stalks (discard the poisonous leaves and green portions). Chop and combine with ½ cup sugar and 1/8 teaspoon cinnamon for every quart of fruit. Let the mixture stand in a cool place, stirring occasionally for about 2-4 hours. Use a juice extractor, or mix 1 quart of water with each 5 quarts of fruit. Bring just to a boil, then strain the fruit and pour the hot juice into hot jars, leaving ½ inch headspace. Process in a boiling-water bath for 20 minutes.

How to Can Vegetables

With the exception of perhaps tomatoes, it is important to remember that vegetables are low-acid, meaning that unless they are pickled, they are extremely vulnerable to harboring deadly bacteria. **Vegetables that are not pickled or lacto-fermented should always be processed using a steam pressure canner.**

Unless otherwise noted, all canned vegetables that are not pickled or lacto-fermented should be hot-packed into jars with 1 inch of headspace in the jar. Most of the processed vegetables also call for saving the liquid they were cooked in to top off the jars (the liquid is filled with nutrients that have leached out of the vegetables while they are being prepared). Use tongs, a basket spoon, or slotted spoon to fill the jars with the vegetables and then pour the liquid in to top off. Processing time is at 10 pounds of pressure, unless otherwise noted.

Asparagus—Spears should be freshly picked, firm, and tightly closed at the tip. Cut to 1 inch shorter than the jar, or cut into 1-inch pieces. Cover with boiling water and gently boil for 2-3 minutes. Pack in jars, tip ends facing up, and add the cooking liquid to the jars leaving 1 inch headspace. Process 30 minutes for pints, 40 minutes for quarts (using widemouth jars makes it easier to remove the spears without crushing them)

Beans, Bush, Snap, and Filet—Trim tips, tails, and strings on tender, freshly harvested beans. Leave whole or cut into 1-inch pieces. Cover with boiling water and boil gently for 5 minutes. Pack in hot jars, add the cooking liquid, leaving 1 inch headspace, and process for 20 minutes for pints, 25 minutes for quarts.

Beans, Fava and Lima—Shell the beans from tender, plump pods. Cover with boiling water and gently boil for 3 minutes. Pack in hot jars and cover with cooking liquid to 1 inch headspace. Process pints for 40 minutes and quarts for 50 minutes.

Beans, Dry and Field-Type Peas—Choose mature, dry beans or peas of uniform size; discard off-color or shriveled beans. Cover with boiling water and gently boil for 3 minutes. Soak beans for 1 hour until they expand to about twice the original volume. Heat to boiling again; drain, reserving cooking liquid. Fill hot jars with hot beans, add cooking liquid, and leave 1½ inches of headspace. I like to add a little salt on the top of the beans (½ to 1 teaspoon, depending on jar size). Process for 75 minutes per pint, 90 minutes per quart.

Beets—Use firm, uniformly shaped (up to 3 inches in diameter) beets. Trim tops and roots to 1 inch to avoid bleeding the color.

Wash and then cover with boiling water and boil until the skins slip off, about 30 minutes, depending on their size. Skin, trim off roots and tops, then pack whole (for small beets), slice, or cube. Pack in hot jars, add 1 teaspoon of salt per quart, if desired, and cover with fresh, boiling water to 1 inch headspace. Process for 30 minutes for pints, 35 minutes for quarts.

Carrots—Use carrots that are sweet, crisp, and no larger than 1¼ inches in diameter. Wash, peel, then rewash and cut into ½-inch-thick sticks or ½-inch-thick slices or chunks. Cover with boiling water and boil gently for 5 minutes. Pack hot carrots into hot jars and cover with cooking liquid, leaving 1 inch headspace. Process for 25 minutes (30 minutes for quarts).

Corn, Cream-Style—Select ears that are perfect for eating fresh. Shuck the corn and gently wash the ears. Drop into boiling

BELOW: At Tunbridge Hill Farm, freshly picked and washed carrots.

water and boil gently for 4 minutes. Cut the kernels from the cob at about ½ their depth, then scrape the cob with a table or butter knife to "milk" the remaining kernels. Don't include any parts of the cob. Heat salted water to boiling in a separate pot (use a ratio of half as much water to kernels and scrapings). Add corn and simmer, stirring gently for about 5 minutes. Pack loosely by ladling mixture into hot jars, leaving 1 inch headspace. Process for 1 hour and 25 minutes in pints only (the mixture is too dense to safely can in larger jars).

Corn, Whole Kernel—Select ears that are ready to eat fresh. Heirloom varieties of corn may be better suited for canning whole; sweeter, hybrid varieties tend to turn brownish (this doesn't indicate spoilage, and it's fine to eat). Shuck the corn and wash gently. Drop in boiling water and boil for 3 minutes. Cut the kernels from the cob at about ¾ their depth (don't scrape the cob or it will cloud the water). In a pot add 1 cup of boiling, salted-to-taste water to every 4 cups of clean kernels and bring to a boil. Simmer the mixture for 5 minutes. Pack the hot jars with the corn and then cover with the cooking liquid, leaving 1 inch headspace. Process for 55 minutes for pints, or 85 minutes for quarts.

Greens—Canning greens such as spinach, mustard, chard, turnip, or kale will turn the leaves soft, so they are suited for use in soups or stews. For firmer texture, freeze greens. Use only crisp, fresh, thick leaves without insect damage. Cut off any roots or tough stems and ribs and drop washed leaves, no more than 1 pound at a time, into a blanching basket or colander in boiling water. Using long tongs, and move the leaves around until well wilted, about 3 to 5 minutes. Loosely drop the leaves into hot jars. Adding ¼ teaspoon each of citric acid (vinegar or lemon juice will do) and salt to the top will preserve the color of the leaves. Add fresh boiling water, leaving 1 inch headspace, and process 70 minutes per pint, 90 minutes per quart.

Okra—Personally, I can't abide the soft texture of cooked okra. On the other hand, after it's pickled, I find it as delicious as dilly beans (see recipes). Use only tender, young pods (if it's not easily sliced through with a knife, it's not tender enough). Wash and leave

Paddle the kitten explores the basement at Fat Rooster Farm, including the shelving that holds the farm's canned goods. From left on the top shelf are canned tomatoes, beets, peaches, and maple syrup.

whole, or slice into 1-inch pieces. Place in a saucepan with hot water and boil gently for 2 minutes. Pack the hot pods, upright if whole, in hot jars and cover with the cooking liquid, leaving 1 inch headspace. Process for 25 minutes for pints and 40 minutes for quarts.

Onions, Small White Boiling—Choose uniformly shaped onions, up to 1 inch in diameter. Peel, trim off tops and roots, and wash gently. Cover with boiling water and boil gently for 5 minutes. Pack the hot onions in hot jars and cover with the cooking liquid, leaving 1 inch headspace. Process for 25 minutes for pints and 30 minutes for quarts. The onions will be soft and sweet; some cooks prefer to brown the onions before serving.

Peas, Edible Pod—Snow peas and other edible pod peas are best frozen, but they can also be canned using the raw-pack method. Trim, wash, and loosely pack the peas into hot jars to 1 inch below the top. Cover with boiling water, leaving 1 inch headspace. Process for 20 minutes for pints and 25 minutes for quarts.

Peas, Green or English—Shell and wash tender, small peas. Cover in boiling water and bring to a boil. Pack the peas loosely in hot jars, about 1 inch from the top, and cover with the cooking liquid, leaving 1 inch headspace (you can add 1 teaspoon of salt per quart jar, if desired). Process for 40 minutes for both pints and quarts.

Peppers, Hot or Sweet—Use crisp, meaty, thick-skinned peppers. **Wear gloves if processing hot peppers.** Wash the peppers, then roast them whole. The easiest way to do this is to place them on a cookie sheet in a hot oven (400 degrees Fahrenheit) or broiler, brushed with olive oil, until their skins begin to blister. If you use a broiler, turn the peppers with tongs as needed, until they're blackened all over. Once the skin has blistered, place them in a covered bowl and let them sit for 10 minutes, or until cool. Scrape off the skins with a knife, then cut off the stem and remove the core, seeds, and inner white membranes. Leave smaller peppers whole, but make two lengthwise slices on each pepper with a sharp knife. Cut the larger peppers into quarters and layer them into hot pint jars or smaller only. Add ¼ teaspoon salt and 1 tablespoon vinegar to each jar, if desired. Add boiling water to the jars, leaving 1 inch headspace. Process for 35 minutes.

ABOVE: Harvested Green Mountain potatoes are set aside for Community Supported Agriculture (CSA) customers at Tunbridge Hill Farm. While the farm's heavy wet soil is great for Allium vegetables (leeks, scallions, onions, and garlic), farmers Jean and Wendy Palthey said they get an average yield for potatoes because they like a soil that is not too wet or too dry.

Potatoes—Canning potatoes renders them firm and sweet, but if they've been stored at less than 45 degrees Fahrenheit, they'll tend to turn brown after canning. Firm, new waxy type potatoes are best for canning. Leave potatoes whole, up to 1 inch in diameter; cut medium potatoes in half. Alternatively, you can cut them into half-inch cubes. Wash the potatoes, peel, and then wash them again. Cook in boiling water for 10 minutes. Pack the potatoes into hot jars, add salt to taste (about 1 teaspoon per quart jar), and cover with fresh boiling water, leaving 1 inch headspace. Process for 35 minutes for pints, 40 minutes for quarts.

· · · · · · · · · · · Pumpkin Soup · · · · · · · · · · ·

Pumpkins are like turkeys; they've both been relegated to limited use during most of the year, except on special occasions like Thanksgiving, when roast turkey and pumpkin pie are the perennial standbys. Many historical recipes used pumpkin instead of winter squash to make everything from bread to pudding. This recipe was adapted from one in *The Joy of Cooking* by Rombauer, Becker, and Becker.

- 2 tablespoons olive oil
- ½ cup chopped onion
- 3 teaspoons minced garlic
- ½ cup chopped celery (including leaves)
- I tablespoon butter
- I tablespoon flour (can substitute any gluten-free flour)
- 3 cups cooked pumpkin
- 3 cups milk, scalded
- ½ cup sour cream
- 2 teaspoons combination of dried herbs, like dill, parsley, lovage, oregano, basil, or sage
- salt and cracked black pepper, to taste

Heat the milk in a soup pot until scalded. Add the cooked pumpkin and blend, then remove from heat.

While the milk is scalding, heat the olive oil and add the onions, garlic, and celery. Cook until soft and translucent but not browned. Add to the pumpkin mixture.

Blend the butter and flour together by kneading it with a spoon and add to the soup.

Add the sour cream, and heat thoroughly, but don't boil. Add the herbs and salt and pepper. Stir well and then puree in a blender until creamy and frothy. Serve hot. Serves 8 as a first course or 4 as the main dish.

ABOVE: From left, the author, Whitney Taylor and Tali Biale peel 10 pounds of garlic cloves to make two cases of Nahm Jeem Gratiem (Thai Crystal Sauce). Minced in a Cuisinart, the garlic is combined with red peppers, sugar, vinegar, water and salt to make the hot sauce that is used as a marinade.

Pumpkins and Winter Squash—Because roasted, mashed, or pureed squash is so dense, it is unsafe to can. Use only cubed pieces of sweet, dry, thick flesh. Wash the whole pumpkins or winter squash and then slice in half. Remove the seeds and stringy flesh, then cut crosswise into 1-inch slices. Peel and cut into 1-inch cubes.

Soybeans, Green—Shell the beans, then cover them with boiling water and bring to a boil. Drain, reserving the cooking liquid. Pack loosely in hot jars and cover with the cooking liquid, leaving 1 inch headspace. Process 40 minutes for pints and 50 minutes for quarts.

Spinach—(see Greens)

RIGHT: Costoluto Genovese tomatoes—an heirloom variety—grown by Fat Rooster Farm are on display for sale at the farmers' market in Norwich, Vermont.

Sweet Potatoes—Wash and sort by size, up to 2 inches in diameter. Boil or steam for 15 to 20 minutes to loosen the skins. Peel, and then pack the hot potatoes, whole or cubed (never mashed or pureed) into hot jars with ½ teaspoon salt. Cover with boiling water or boiling syrup (3/4 cup sugar dissolved in enough water to make a quart and brought to a boil is enough for 5 pints of potatoes), leaving 1 inch headspace. Process for 65 minutes in pints if cubed, 90 minutes in quart jars if using whole tubers.

Turnips—Young turnips are sweet and crisp after canning. Use tender turnips, no larger than 3 inches in diameter. Wash and peel. Cube or slice and boil gently for 5 minutes. Pack hot turnips into hot jars and add the cooking liquid, leaving 1 inch headspace. Process 30 minutes for pints and 35 minutes for quarts.

How to Can Miscellaneous Foods

Mushrooms—Almost all food preservation books warn against canning anything but domestic mushrooms. I believe this is because of the potential for misidentification and then intensifying the adverse, sometimes deadly effects of the button mushroom's wild cousin. If you are a proficient mushroom hunter and sure of your identification skills, play it safe, and pickle wild mushrooms (see page 36). For domestic mushrooms, use fresh, unblemished mushrooms with firmly closed caps. Trim off the stems up to the caps (use these in a gravy or soup), and soak caps in cold water for about 10 minutes. Then rinse them in fresh water. Cut larger mushrooms in half; leave smaller ones whole. Cover the mushrooms in cold water, bring to a boil, and simmer for 5 minutes. Pack the hot mushrooms in hot jars and add ⅛ teaspoon of ascorbic acid or one crushed 500-mg vitamin C tablet and ¼ teaspoon of salt per pint.

Add fresh boiling water leaving 1 inch headspace. Process only half pints or pints for 45 minutes (mushrooms are the king of low-acid foods, and processing quarts would not adequately remove harmful bacteria).

Tomatoes—Most home canners use paste varieties of tomatoes for canning instead of slicers; their flesh is firmer, and they tend to have less seeds and juice. I use a variety of tomatoes to create more flavor in the sauce. You can always take out the majority of seeds before the tomatoes are canned. Another processing method that is gaining in popularity is using a vegetable mill or strainer. These manually operated machines separate out pulp and juice from skin and seeds without having to first peel or remove the seeds.

Canning tomatoes whole or halved allows you to preserve the food quickly when there's a real surplus and think about how to use it later; a jar of stewed tomatoes can always be turned into sauce or salsa later. The USDA also strongly warns against adding other low-acid foods to the sauce without following a specific recipe due to bacterial concerns (some deadly bacteria, such as *Clostridium botulinum*, thrive in low-acid, low-oxygen environments). If you'd rather be creative with your sauce and add whatever is available in the garden, you can always freeze it.

On the downside of canning tomatoes whole or halved is the fact that the jars take up more room than processed sauce or ketchup does. Alternatively, you could crush the tomatoes and can them, condensing the jars to some degree (see below). Using a food mill to puree the fruits will greatly reduce the time and wasted space in jars.

The USDA recommends adding 2 tablespoons of lemon juice per quart jar because tomatoes are considered borderline low-acid, especially heirloom varieties and overly ripe fruits. Vinegar is a suitable alternative—4 tablespoons per quart. Many home canners also add salt to taste.

While the tomatoes are being prepared for canning, I sterilize the jars in the hot-water bath. I like to use quart jars, preferably widemouthed, for easier filling. To prepare the tomatoes for canning without a food mill or strainer, the skins should be removed, or the end product will have stringy skins throughout. Wash the tomatoes and then plunge them in a pot of boiling water to loosen their skins, about 15-30 seconds, depending upon how ripe they are (ripe ones take less time, as do heirloom slicers or cherries versus thick-skinned paste varieties). I use a 10-quart pot and dump 2-4 pounds of fruits in at a time, depending on their size. Remove them from the pot with a basket or slotted spoon and dip them in cold water. The skins should easily peel off, using a paring knife. The tomatoes are then ready to be quartered and crushed or canned halved or whole.

I don't like to add additional liquid to top off filled jars, so I pack the raw tomatoes firmly in the hot, sterilized jars, pressing them until their juice fills the spaces between them. I then add the lemon juice and salt, leaving ½ inch headspace. Technically, this

Morning dew dissipates from tomato plants at Nicola Smith's home. Smith rotated the tomato plants from their usual garden spot to an area where the family's half-dozen chickens spent the winter. The plants grew well and weren't affected by the tomato blight.

is a raw-pack method, but the tomatoes are often warmed from the skinning procedure. The lids are adjusted, and then they're processed in a hot-water bath for 45 minutes for pints and quarts alike (the USDA recommends 85 minutes).

A pressure canner can also be used and will result in a more nutritional end product. Quart jars should be processed at 11 pounds of pressure for 25 minutes in a dial-gauge canner and at 15 pounds of pressure for 15 minutes for a weighted-gauge canner.

If you're using tomatoes that have been crushed and pureed using a food mill or strainer, heat the mixture to boiling and gently cook for 5 minutes. Fill the jars (preferably quarts), add the acidifier and salt to taste, leaving ½ inch headspace. Process in a hot-water bath for 45 minutes or in a pressure canner for 15 minutes.

Tomato Juice—Wash the fruits, cut out the stems, and chop all but 1 pound of the tomatoes. Slice the remaining tomatoes into

quarters over a plate to catch their juices. Put these into a pot and heat to boiling while crushing with a potato masher. Slowly add the remaining tomatoes to the sliced fruit, making sure the mixture is always boiling, and crushing the tomatoes as you go. Simmer the tomatoes for 5 minutes after all the fruit has been added, then press the mixture through a sieve or food mill and bring the strained juice to boiling.

Add 2 tablespoons lemon juice and 1 teaspoon salt (if desired) to each hot quart jar and fill with the hot liquid, leaving ½ inch headspace. Process in a boiling water bath for 40 minutes or in a pressure canner for 15 minutes.

To prepare a tomato juice with a kick, you can add up to 1½ cups of other chopped vegetables such as carrots, onions, celery, or sweet or hot peppers to every 12 pounds of tomatoes. Simply follow the steps for plain tomato juice, and when all the tomatoes have

been crushed, add the extra vegetables. Simmer this mix for 20 minutes, stirring as you simmer. Strain the mixture through a sieve or food mill. Pour the liquid into hot jars, and add 2 tablespoons lemon juice and 1 teaspoon salt (optional) per quart. Process in a boiling-water bath for 35 minutes for pints, 40 minutes for quarts. Alternatively, you can process in a pressure canner for 15 minutes.

LEFT: From left, kale, arugula and red mustard is planted together for a spicy green mix that is harvested together for salad greens at Fat Rooster Farm.

Jars of "nontraditional" kimchi, with their lids on but not sealed, are ready to be eaten in Donna Kausen's kitchen. Kausen said she made it without cabbage, but included turnip, celeriac, and hot pepper. "But you can make it out of anything," she said.

" Where once beer, bread, and cheese were quirky local products varying from place to place, we lucky twenty-first-century consumers can buy fermented commodities such as Bud Lite, Wonder Bread, or Velveeta that look and taste the same everywhere. Mass production and mass marketing demand uniformity. Local identity, culture, and taste are subsumed by the ever-diminishing lowest common denominator, as McDonald's, Coca-Cola and other corporate behemoths permeate minds on a global scale to create desire for their products.**"**

—WILD FERMENTATION: THE FLAVOR, NUTRITION, AND CRAFT OF LIVE-CULTURE FOODS, *Sandor Ellix Katz*

⑤

Pickling and Fermentation

THE ART OF PICKLING foods can be traced back thousands of years, to the very first wines made of honey, created by ancient cultures in India, Spain, and South Africa. Today, pickling most commonly refers to a method of food preservation that usually involves either salt or a strong acid being introduced to the food in its original state, but natural fermentation is also possible. Living microorganisms, such as yeasts, molds, and bacteria that occur wherever we live can change the composition of the food by creating an acidic or alcoholic environment that is inhospitable to the bacteria that cause food poisoning. This process often makes the end product more nutritious, more digestible, or transformed from something that does not keep well without spoiling (milk or soybeans) to an end product with superior keeping qualities (cheese or tamari).

Most of us are familiar with traditional pickles like dilly beans or cucumber pickles, where cultured vinegars and salt are added to "brine" and "cure" the raw vegetables. Sauerkraut is also a food that is generally recognized as fermented, having been shredded, salted, and then left to ferment in a crock or food-grade plastic bucket until it's ready to can and later serve with salty meats like ham. However, foods such as chocolate, coffee, tea, and vanilla are also pickled or fermented before the final product is consumed. The ancient Romans made *liquamen* from fishes that were packed in salt and fermented for several months. The resulting amber-colored liquid was so common that almost no meal went without its addition, referred to in present day as fish sauce, or *Nam pla* in Thailand, and *Nuoc nam* in Vietnam.

One of the oldest forms of pickling is called lactic fermentation (or lacto-fermentation). In vegetables, the starches and sugars are converted into lactic acid by lactobacilli—naturally occurring bacteria in living tissue, and most abundant in leaves and roots of ground-living plants. Besides the ubiquitous sauerkraut, lacto-fermented foods are important in cultures worldwide. The Swedes have a dish called lutefisk, which is dried cod that has been steeped in lye, then fermented for several weeks before it's rinsed and cooked. The dish is part of a traditional Christmas Eve feast, along with several other fermented vegetables and meats. In Asian countries, fermented foods such as kimchi, sushi, and fermented vegetables like turnip, radish, onion, cabbage, and carrot are part of the daily meal. In the United States, relishes and pickles were most likely lacto-fermented prior to the modern-day practice of adding vinegar and salt to preserve (after all, vinegar is a fermentation by-product).

Lacto-fermentation may have fallen out of favor due to the fact that industrialization of products preserved by this method is not feasible; each outcome is largely determined by the yeasts, molds, and bacteria that reside in a specific region, thereby creating conditions that can produce artisanal, regional foods, not mass-produced homogenous ones. Indeed, the famous San Francisco sourdough relies on a lactobacillus found only in that part of the country.

Pickling (using a salt and vinegar solution to preserve) has become much more popular because of its consistent results. However, the final product is much more acidic, which may not be healthy if consumed in great quantities; the act of pasteurizing the food to preserve it in sealed jars kills the bacteria responsible for producing lactic acid, and in the end deprives the consumer of the added digestibility of foods preserved with these bacteria.

In this chapter, simple pickling and lacto-fermentation of fruits and vegetables will be explained to get you familiar with these preservation techniques.

Preserving Through Lacto-fermentation

To successfully preserve foods using this technique, there are a few basic rules:

- Use only fresh, vibrantly healthy vegetables. These will contain the largest amounts of naturally occurring lacto-bacilli, responsible for fermentation and preservation.
- Use food-grade plastics or glass containers in which to ferment your foods. Never use reactive containers such as metal. Widemouthed glass jars are ideal for lacto-fermenting.
- Don't use chlorinated tap water—it will kill off the bacteria responsible for fermentation.
- Use only sea salt or kosher salt, not iodized table salt.
- When fermenting dairy products, seek out a source for fresh, raw milk that contains living bacteria that have not been killed through pasteurization.
- If you're using whey to jump-start the lacto-fermentation process, be sure to culture the whey at least 3 days before proceeding with the rest of the recipe.

Adding Whey to Facilitate Fermentation

Author Sally Fallon in Nourishing Traditions (1999) advocates using a small amount of whey when lacto-fermenting. Whey is a natural by-product of cheese making; it's the watery part that is left over after the curds have been separated and drained. While it's not

Kyle Jones runs a rototiller over soil where cloves of softneck garlic are to be planted at Fat Rooster Farm. The heavy, clay soil needs to be worked over

· · · · · · · · · · · · Mama's Dough- · · · · · · · · · · · Pickles—transcribed by Aunt Anita Megyesi, 2002

- Widemouthed gallon glass jar
- 8 tender, fresh-picked grape leaves
- 6 dill heads (including some of the stem)
- 12 cloves garlic
- approximately 5 pounds fresh pickling cukes, 3–5 inches long
- ½ cup kosher salt (do not use iodized salt)
- 1 quart water
- 2–4 slices homemade rye or pumpernickel (do not use store-bought, soft, white bread)

Layer the bottom of the jar with half the grape leaves, then half the dill heads and the garlic cloves.

Cut off the blossom end of the cukes, scrub, and rinse with cold water. Starting from one end, cut each cuke into quarters, to three quarters of its length (don't cut through). Pack the cukes in the jar, intact end first.

Make the brine by dissolving the salt in the water. Pour the solution over the cukes, covering them. Add more brine if needed.

Top off the cukes with another layer of grape leaves, dill heads, and garlic cloves. Place the bread slices on top of the pickles and cover the mouth with cheesecloth secured with string or a rubber band to keep the muslinca (tiny flies) out.

Keep the pickles at room temperature in a dark place for 4–6 days. They'll turn olive drab to light yellow in color, and the texture will turn soft and crisp. They're ready to eat at this point—overfermentation will make the cukes too soft and too punchy.

Store pickles packed in quart jars with strained brine for up to 2 weeks.

very tasty (in the Hebrides Islands off Scotland, the drink was historically called "bland"), it contains about a third of the milk's original proteins and all of its sugars. It's also packed with lacto-bacilli, so adding it to the vegetables being preserved will bring more consistent fermenting results instead of relying on the natu-rally occurring bacteria that are present in widely varying amounts within the vegetables.

If you have a source of raw, whole milk, you won't need a starter culture. Just leave the milk out in a glass container at room temper-ature until the milk separates, about 2 days. The end product should smell sweet or slightly sour, but not spoiled.

After it has separated, the whey needs to be drained from the rest of the milk solids. You can do this by layering a sieve or strainer with cheesecloth or cheese linens and pouring in the milk. Wait for several hours (I like to pour off the cultured milk early in the morning and come back to the project around lunchtime), then carefully tie the cloth or linen to form a ball. Don't squeeze the solids. Tie it to a wooden spoon or dowel and allow it to drain until it no longer drips. The solids can be used like cream cheese, and the liquid whey will keep in a glass jar for up to 6 months if kept below 40 degrees Fahrenheit.

Using Salt in Your Fermented Foods

The amount of salt recommended for preventing the growth of pu-trefying bacteria varies widely. The more salt added, the slower the fermentation process becomes, and with too much salt, the process will be halted altogether.

Older recipes suggest brining vegetables such as root crops, cabbage, brussels sprouts, and broccoli in a 10 percent brine solu-tion, where a fresh egg will freely float on the surface of the brine. Still others suggest that no salt at all is needed to properly lacto-ferment.

Generally, 1–2 tablespoons of salt per quart of water will make sufficiently salty brine. If no water is added, use ³/₄ to 1½ teaspoons of salt per pound of vegetables. The brine should be sufficiently salty, but not strong enough to make it seem like you're tasting seawater. Keep a log of your results and adjust the saltiness the next time.

Lacto-fermentation requires anaerobic conditions, meaning that oxygen should not come in direct content with the food being

fermented. If you're using food-grade plastic buckets or crocks, the vegetables need to be weighted down so that they're covered by their juices. A jug filled with water set upon a plate will do the trick, as will a clean stone or other nonreactive weight. The food can also be fermented directly in widemouthed canning jars that are sealed and kept at room temperature for 3 or 4 days before storing in a cool, dark place.

LEFT: Jean Palthey picks broccoli for market at Tunbridge Hill Farm. The Paltheys believe their vegetables' unique taste is due to the quality of the soil on their farm, which was a hillside dairy farm for years.

Bright Lights swiss chard leaves are cut at the base, but grow back. Baby leaves are best in salad, and fully-grown leaves can be frozen for later use.

6

Freezing Fruits and Vegetables

EVERY SUMMER, I CAJOLE *someone on the farm into seeking out one of my favorite wild foods with me: wild mushrooms.*

Usually, their abundance coincides with the most intensely busy times of the year: morels come when the first farmers' markets have started and the greenhouse is full of plants that need tending to. Chanterelles come at the height of summer, when all we think of is getting hay in the barn. And my favorite, chicken of the woods, falls in between, with nary a specified date, dependant on rain and heat to grow into colossally huge arrays of orange, succulent fungi whose best attribute is that they're great even after being frozen.

Mushroom foraging is a luxury, one right up there with having the bed made and the lawn mowed. It's an extravagance, because, honestly, who ever heard of surviving the winter or even gauging one's readiness for the bare season that follows summer's bounty by the amount of mushrooms lardered away? There are several pounds of beans to pickle, beets and carrots to store, and Swiss chard to bunch and freeze before even the most delectable of mushrooms can be considered.

Today, I have been enticed away from the farm by my sister's boastings of a mushroom field that even I will not scoff at. She says that she's harvested 3 pounds of mushrooms and not even dented its surface.

As a family, we tend to exaggerate, so I am not prone to believe her. We walk through the woods where I used to wander as a fourteen-year-old, a place safe enough that our parents would not think to try and find us until

What to Do if You Lose Electricity

First of all, determine whether all of the power is out, or just your freezer has quit. If you have another freezer that is working, transfer the food immediately, and turn the freezer to its coldest setting.

If the failure is due to a power outage, a fully stocked freezer will keep its contents frozen for as long as 2 days, provided you don't open up the door! A half-loaded freezer may only keep food frozen for a day, but freezers kept in cooler spaces will keep food frozen longer.

If you've determined that your loss of power will go on longer than a couple of days, you can pack the freezer with cardboard and paper, purchase dry ice to put on top of the food, or remove the food, cook it, and refreeze it soon after (cooked food will not last as long in the freezer).

After you've resumed power, check the packages. Turn the freezer setting to its coldest, and mark any partially thawed packages so that they're consumed first. Don't taste or eat any room-temperature foods that have been warm for longer than 2 hours; they can smell fine but still contain deadly bacteria. Make sure not to let pets or livestock eat the spoiled food; compost it and use it on the garden after it's well rotted and no longer resembles the original food.

ABOVE: Frozen black currants are stored in a chest freezer in the former milkhouse at Fat Rooster Farm.

dusk set in. Now, there are new houses where horse trails were, but still, it's not as bad as I expected (I've actually refused to walk in these woods for the past eleven years, in fear of what I would find. My sister has finally told me that most of it is still there, the tree shaped like the number 4, the sliding hill, the moss rock). One of the houses is beautifully done. It looks like an old New England saltbox, complete with grey stain, a roof that resembles cedar shakes, and a horse barn finished with old-fashioned windows.

We walk by these places, and I remember what they looked like, devoid of houses, when I rode on my horse by them many years ago, maybe twenty years ago, maybe thirty.

The sky is breaking clear blue, and I think about how different the weather is, right now, at this moment, just 50 miles away. I think about the apprentices, diligently weeding the vegetable fields despite the clouds and dampness, trying to stay on top of it, allowing me to escape and take advantage of this rainy weather and all the mushrooms it has brought. Then we get to the chanterelles.

There is no way that I can describe what this looks like. There are orange carpets of mushrooms as far as we can see. Everywhere. We start picking, and despite the temptation to be gluttonous, we're choosy, careful not to take the old ones who've sent their spores, or the very young ones who'll continue to produce. There are other mushrooms mixed in among the chanterelles: frilly coral mushrooms, little ones with bright green caps, burnt orange, sun-yellow, velvet brown. It's like we're in a terrarium.

Back home, after absconding with 11½ pounds of mushrooms, we are greeted with the news that we have the tomato blight—the one that has been spread by big-box stores selling to home gardeners who have the desire to grow and preserve their own food, and who have unknowingly purchased and spread the disease—the one spread by a fungus called Phytopthera infestans. What a rough justice that I have spent these wee bits of time harvesting fungi, when a fungus has destroyed our tomatoes.

We pull up and burn 300 of the 600 plants, hoping that the remaining ones don't die. Shannon swears, and we all just look, having only heard her grunt in discontent in the past. She planted the seeds, transplanted the plants in their pots, planted them in the ground after making the holes in the biodegradable plastic and mulching the rows with hay. Staked them, tied them, watched them grow fruits. Today, she ripped them out of the ground and burned them. Guess it's worth the word.

Tali quotes Godfather lines, trying to keep it light. We all work like mad, to get rid of the spores, to stop it, to protect what's left. All that we have read points to total failure, but right now, we're up to accepting this horror, not that one.

At supper, we feast on the chanterelles that would not be so plentiful if not for the rain. We don't talk about the pitfalls. We just rejoice in what we've found in this wet, cool summer—like chicken of the woods.

History of Refrigeration

Human beings have used refrigeration and freezing as a preservation technique for millennia. Ancient caves and dugouts were used to keep food reserves cold during warmer months by lining the dwellings with plant material and other insulation and then packing them with ice.

Toot the farm cat keeps an eye on activity at Fat Rooster Farm. Abandoned as a kitten at a nearby general store, Toot was nursed him back to health by the author.

In more modern times, springhouses were common, though their purpose was slightly different. Here, a building stood over an existing source of water, or spring, and perishable items like meats and dairy were placed in crocks in a trough in the floor where water coursed through. Shelves built above the trough could also maintain cool, dry conditions for storing fruits and vegetables, though these shelves were too damp and humid for many storage crops like onions and winter squash.

Ice houses were also important for storing food as well as an important part of the economy of New England. Ice from frozen ponds in Massachusetts, New Hampshire, and Vermont was harvested and packed in straw or shavings. The ice was shipped as far as New Orleans and parts of the Caribbean, even across the Atlantic to India. Food could be stored throughout the hotter months of the year, but ice was expensive, and only the elite could afford to keep it throughout the year.

Affluent families and city dwellers had iceboxes situated directly in their homes to keep food cool. They allowed people to move to largely populated cities, far away from the farm, and thus the food source.

Frozen tomatoes are stored in a chest freezer in the former milkhouse at Fat Rooster Farm.

Iceboxes were made of oak or walnut and roughly resembled the modern-day refrigerator, only they were shorter. Iceboxes were usually lined with tin and insulated with cork, seaweed, or straw. Ice blocks, delivered directly to the house by the iceman, kept perishables cold.

After reigning for over one and a half centuries as the preferred and most efficient form of refrigeration, the icebox was dethroned by the mechanized, chemical refrigerator. These were first introduced in households during the early 1900s, but the coolants were dangerous, being both flammable and toxic. In the 1920s, it cost 1/3 more to purchase a refrigerator than it did to own a new Model-T Ford. By the 1930s, the prices had fallen drastically, and the popularity of refrigeration soared. Today, it's not uncommon to be able

RIGHT: Even while it's snowing at the end of December, kale can still be harvested from the frozen ground at Fat Rooster Farm.

to purchase a freezer for less than it costs to drive to see the grand-parents in Ohio from Vermont.

Benefits of Freezing

Hands down, freezing requires less time to preserve the harvest than any other method available. You simply harvest, prepare, blanch, and pack into containers, and the food is ready to cook without having spent hours drying, cooking, and canning before the food can be prepared for eating.

Not only is it easy to do, freezing your food will result in the least amount of nutrient loss, particularly of water-soluble vitamins such as the vitamin C contained in leafy greens, root vegetables, and fruits.

Using freezing as a technique to preserve your harvest is not foolproof. Only the highest quality foods should be frozen, or they'll come back out tasteless and watery, resembling cardboard in texture. Food preparation for freezing should be done quickly and carefully, and the containers you use should keep moisture and air out. If the frozen foods aren't kept at a constant tempera-ture, ideally below o degrees Fahrenheit, they'll become freezer burned—riddled with tiny ice crystals that suck the moisture out of food and reduce its quality.

How Freezing Works

Food is frozen through crystallization. Water surrounding the out-side of the food, between the food cells, freezes and forms crystals first. The water within the food cells is then pulled out of the cells and frozen, leaving the cells concentrated in salts and enzymes. If this process occurs quickly, the formed crystals will be small, not jagged and long, and the cell walls of the food won't be harmed. If the freezing time is prolonged, or the foods are allowed to freeze and thaw, the crystals will damage the food's cell walls and damage its structure and texture.

Keep in mind that freezing doesn't kill microorganisms like canning does, so bacteria, molds, yeasts, and parasites are still present, just "frozen" in a state of limbo. Even enzymes are only slowed in their activity unless the food is first boiled (or blanched), and changes in the food's color and loss of its nutritional value will eventually result.

Equipment You'll Need for Freezing

When purchasing your freezer, you should keep the following in mind:

- energy efficiency
- amount of space needed for frozen foods
- ease of loading and unloading frozen goods

House guest Wiggy, a terrier mix, is let outside at Donna Kausen's home. Kausen and her late husband, Bradford, built the home between 1990 and 1993 with Balsam Fir logs they cut on their property. At left is an unheated pantry used for food storage.

After 1993, all freezers were required to pass energy efficiency standards. Purchasing a used freezer is always an option, but be careful to inspect the seals around the freezer lid; replacing these seals can cost sometimes nearly a third of the total cost to replace the freezer. Also inspect it both inside and out, looking for dings, off odors, and other defects. Be sure that the freezer will maintain temperatures that are 0 degrees Fahrenheit or below.

Generally speaking, you should allow about 6 cubic feet per person in the family; 1 cubic foot will hold about 35 pounds of frozen food.

The decision to purchase a chest versus upright freezer will depend on how you intend to use it. Chest freezers hold more food per cubic foot, and they lose much less cold air when they're opened. They're generally less expensive to purchase, and because they're more efficient at staying colder, they'll be less costly to run.

When Not to Use Freezing for Preserving Your Food

Some foods just don't do well being frozen. These include vegetables like green bunching onions, many salad greens, radishes, and tomatoes (unless they're frozen whole and used for soups and stews later or they're frozen as cooked sauce). Root cellars are often better at preserving the crunchy texture and sweetness of many vegetables like Jerusalem artichokes, cabbage, celeriac, and even carrots. Obviously, freezing food offers the convenience of it being already prepared to eat after warming, but if you have the ability to cellar some of the crop, make sure to take advantage of this superior storage technique.

Some disadvantages with a chest freezer are that it's harder to take food out (you have to reach down into it and pull the food out) and they're harder to organize. Foods can quickly become buried unless baskets and specific containers are used within it.

Organizing Your Freezer

As with all food that is preserved, each container or package within the freezer should be labeled with the date and contents. Use a Sharpie-type permanent marker, not pencil or a marker that could bleed through and taint the frozen food within.

When you put foods in the freezer, be sure that the most recently added containers or packages are on the bottom, so that you use the food you've frozen the longest first.

Make sure that foods are cooled sufficiently before they're frozen, or the insides of the package may not freeze fully and will spoil. Fruits and vegetables can be prepared by washing, trimming, and peeling if needed, then cut into pieces for blanching

BELOW Frozen corn is stored in a chest freezer in the former milkhouse at Fat Rooster Farm.

Using Frozen Fruits and Vegetables

You can use a microwave to blanch foods prior to drying and freezing them. Advantages are that more water-soluable vitamins can be retained, flavor can be preserved better by microwaving, you won't heat or steam up your kitchen as much using a microwave. Still, you can quickly cook foods too much using a microwave, making them mushy and less palatable and microwaving won't inactivate enzymes as well as boiling water blanching, so freezer storage time can be reduced.

(if needed). Quickly cool blanched foods in cold water, drain, and then lay the foods in a single layer on trays to cool. Foods intended for freezing should be packed in the freezer as quickly as possible; don't stack more than more than 3 to 5 pounds of new food per cubic foot of freezer space in more than a day; the food on top will freeze, but the unfrozen food can rot before it has a chance to fully freeze. The freezer should be maintained at its coldest possible setting until the foods have completely frozen.

There are several materials that are suitable for storing foods in the refrigerator. Perhaps the easiest and most reliable of these are rigid plastic freezer containers and plastic freezer bags. Make sure that they're labeled for use in the freezer—don't use garbage or lawn bags, as the plastic could impart off-flavors or even toxic chemicals on the foods.

I prefer freezer bags, particularly for fruits and vegetables, because they can be laid flat to freeze quickly, and then stored in neat stacks.

Empty plastic yogurt, cottage cheese, and ricotta containers are also handy, particularly for storing liquids like soups and broths.

For foods that are baked, precooked, or prone to drying out, heavy-duty aluminum foil or freezer paper are excellent packaging materials, although aluminum foil can tear easily if the foods are

Table 4 Part One

FRUITS	Fresh (lbs)	Frozen (pts)	Freezer Lifespan (months)	Prepare Fruit By *	
Apples	1 1/2	1	6	core, slice 1/4" thick	
Apricots	2/3	1	6	blanch first, halve, pit, slice	
Bananas	varies	varies	1	whole, unpeeled	
Blackberries	1 1/2 pts	1	6	whole, unwashed for dry-pack	
Blueberries	1 1/2 pts	1	6	whole, unwashed for dry-pack	
Boysenberries	1 1/2 pts	1	6	whole, unwashed for dry-pack	
Cherries (Sour)	1 1/2	1	6	stem and pit	
Cherriess (Sweet)	1 1/2	1	6	stem and pit	
Coconut	varies	varies	3	shred, chop or grate	
Cranberries	1/2	1	6	whole, unwashed for dry-pack	
Currants (Dried)	varies	varies	6	whole	
Currants (Fresh)	3/4	1	6	whole, unwashed for dry-pack	
Dates	varies	varies	6	split, remove pits	
Dried Fruits	varies	varies	12		
Elderberries	1 1/3	1	6	whole, unwashed for dry-pack	

* All fruits should be washed prior to freezing unless noted
** Refer to Page 141 for Syrup Recipes
*** Refer to Page 140 for Antibrowning Recipes

Blanch	Dry-Pack	Sugar Pack (sugar to fruit in cups)	Syrup Pack (%) **	Antibrowning Agent (tsp) ***
yes, 2 min, or peel	yes	1/2 c to 4 c	40%	1/2
yes, 30 seconds, or peel	yes	1/2 c to 4 c	40	3/4
no	yes	no	no	n/a
no	yes	3/4 c to 4 c	20 to 40	n/a
no	yes	1/2 c to 4 c	20 to 40	n/a
no	yes	3/4 c to 4 c	20 to 40	n/a
no	yes	3/4 c to 4 c	50	1/2
no	yes	2/3 c to 4 c	40	1/2
no	yes	1/2 c to 4 c	no	n/a
no	yes	no	50	n/a
no	yes	no	no	n/a
no	yes	3/4 c to 4 c	50	n/a
no	yes	no	no	n/a
no	yes	no	no	n/a
no	yes	3/4 c to 4 c	20 to 40	n/a

Continued...

FRUITS	Fresh (lbs)	Frozen (pts)	Freezer Lifespan (months)	Prepare Fruit By *	
Figs	varies	varies	6	whole	
Galangal and Ginger	varies	varies	2 weeks	whole	
Gooseberries	1 1/2 pts	1	6	whole, unwashed for dry-pack	
Grapes	varies	varies	6	whole	
Guavas	varies	varies	6	whole, unpeeled	
Kiwis	varies	varies	6	peel, slice 1/4 inch thick	
Lemons and Limes	varies	varies	6	whole	
Mangoes	varies	varies	6	peel, pit, slice	
Melons	1 1/4	1	6	remove rind, seeds, cube	
Papayas	varies	varies	6	peel and slice	
Peaches	1 1/2	1	6	remove pit, quarter or slice	
Pears	1 1/4	1	6	peel, halve, quarter or slice	
Persimmons	varies	varies	6	whole	
Pineapple	1 1/4	1	6	peel, core, cut as desired	
Plums	1 1/2	1	6	whole or halve, pit and cut	
Raisins	varies	varies	6	whole	
Raspberries	1 pt	1	6	whole, unwashed for dry-pack	
Rhubarb	1	1	6	trim stalks, cut	
Strawberries	2/3 qt	1	6	remove hulls, whole or slice	
Sweet Citrus Fruits	varies	varies	6	peel, slice into segments	

Blanch	Dry-Pack	Sugar Pack (sugar to fruit in cups)	Syrup Pack (%) **	Antibrowning Agent (tsp) ***
peel if desired	yes	3/4 c to 4 c	40	n/a
no	yes	no	no	n/a
no	yes	no	no	n/a
no	yes	no	40	n/a
no	yes	no	no	n/a
no	yes	no	no	n/a
no	yes	no	no	n/a
no	yes	1/3 c to 1 c	30	n/a
no	yes	no	20	1/4
no	yes	no	20	n/a
yes or peel	no	1/3 c to 4 c	20	1
no	no	no	40	3/4
no	yes	no	no	n/a
no	yes	no	no	n/a
no	yes	no	50	1/2
no	yes	no	no	n/a
no	yes	3/4 c to 4 c	40	n/a
no	yes	no	40	n/a
no	yes	3/4 c to 4 c	50	n/a
no	yes	no	40	n/a

Table 4 Part Two

VEGETABLES	Fresh (lbs)	Frozen (pts)	Freezer Lifespan (months)	
Artichokes (Globe)	varies	varies	4	
Asparagus	1 1/2	1	4	
Avocados *	varies	varies	2	
Beans, Dry and Field Peas	varies	varies	12	
Beans, Lima and Shell	2 1/2	1	4	
Beans, Snap	1	1	4	
Beans, Soy and Fava	2 1/2	1	4	
Beets	1 1/2	1	6	
Broccoli	1	1	6	
Brussels Sprouts	1	1	6	
Cabbage, Chinese and Head	varies	varies	4	
Carrots, Pieces	1 1/2	1	6	
Carrots, Whole	1 1/2	1	6	
Cauliflower	1 1/3	1	6	
Celery	varies	varies	6	
Corn, Sweet, in Husks	2 1/2	1	6	
Corn, Sweet, Whole Ears	2 1/2	1	6	
Corn, Sweet, Cream-Style	2	1	6	
Greens (All Types)	1 1/2	1	4	
Kale	1 1/2	1	4	
Kohlrabi	varies	varies	6	
Leeks	varies	varies	6	

Prepare Vegetables By	Blanch (minutes)
trim leaves down to heart	7
peel tough stalks, cut or leave whole	2 to 4
peel, pit and puree	no
do not wash	no
shell	2 to 4
trim, string, leave whole	3
use only tender pods	4 to 5
leave whole, unpeeled, cook fully then peel	no
peel tough stalks, cut into bite-sized pieces	3
leave whole or split larger sprouts in half	3 to 4
cut into wedges or shred	1 1/2 to 3
wash, peel and cut	2
wash and peel	5
cut into bite-sized pieces	3
cut stalks into 1/4 " slices	2
remove silks	no
remove husks, silks, trim ends and rinse	7 to 9
remove husks, silks, blanch, cut kernels from cob	4
wash, remove tough ribs, roughly chop after blanching	2 to 3
prepare as for Greens	3
wash, trim tops and stems, dice or cube	2
clean (white part only) thoroughly then chop	no

Continued...

VEGETABLES	Fresh (lbs)	Frozen (pts)	Freezer Lifespan (months)	
Mushrooms *	varies	varies	4	
Okra	varies	varies	6	
Onions	varies	varies	6	
Parsnips, Salsify and Scorzonera	1 1/2	1	6	
Peas, Snap and Snow	2 1/2	1	6	
Peas, Green	2 1/2	1	6	
Peppers, Hot	varies	varies	6	
Peppers, Sweet	2/3	1	6	
Potatoes	varies	varies	6	
Rutabaga	1 1/2	1	4	
Shallots	varies	varies	1	
Squash, Summer	varies	varies	6	
Squash, Winter and Pumpkin	varies	varies	6	
Swiss Chard	1 1/2	1	4	
Tomatillos	varies	varies	6	
Tomatoes, Green	varies	varies	4	
Tomatoes, Ripe	varies	varies	1	
Turnips	1 1/2	1	4	

* Add 1 Tbs lemon juce for every two avocados to prevent darkening; for mushrooms, soak in 2 tsps lemon juice

Prepare Vegetables By	Blanch (minutes)
riinse, trim, pat dry	3 to 5
wash, leave whole	3 to 5
wash, cut in 1/4" slices or chop	no
wash, trim, slice or cut into cubes	2
wash, trim tips and tails, leave whole	1 1/2 to 2
wash pods, then shell	2 to 3
trim tops, roast then peel	no
trim tops, halve, core and seed (can roast and peel)	2 to 3
bake or fry; don't blanch or freeze uncooked	no
wash, peel , fully cook and mash for best flavor	no
prepars as for Onions	no
cut into 1/2" thick slices or shred then blanch	3
wash, halve, roast, seed and mash or puree	no
prepare as for Greens	2 to 3
remove papery husks, cook until tender	no
wash, core, cut into 1/4 " thick slice	no
leave whole, unpeeled, or cook and puree after peeling	no
wash, trim, slice or cut into cubes	fully cook

jostled about repeatedly. Plastic wrap can also be used, provided it's intended for freezer use.

Widemouthed glass jars can also be used, though I tend to stay away from these breakable containers. Also, if they're placed in boiling water for later use in canning, or in the microwave, they can shatter and break.

Because freezing only renders microorganisms contained within foods in a state of suspended animation, when the foods are thawed, they're more perishable than had they been fresh. The instant the foods are thawed, the microorganisms will spring back into action, breaking down and ultimately spoiling the food. This is why it's not recommended to thaw foods at room temperature—the outer edges will thaw, and the microbes will begin degeneration of the food while the middle of the package is still frozen. For the safest practice, thaw foods at temperatures less than 40 degrees Fahrenheit, ideally in a refrigerator.

Many foods can be thawed during the cooking process, particularly fruits and vegetables. Cooking them directly will also reduce the amount of liquid escaping from the damaged cells so it remains more nutritious, tastier, and not watery.

Refreezing Thawed Foods

Whenever you refreeze thawed foods, the quality—taste, texture and nutritional content—will be compromised. However, you can safely refreeze foods if:

- they feel refrigerator-cold (40 degrees Fahrenheit or colder)
- they still contain ice crystals
- they have been thawed and subsequently cooked
- they have been purchased as commercially frozen foods, have thawed and then are refrozen (commercially frozen foods are much more quickly frozen, so they food's cells are not damaged as much as home frozen foods are).

Never refreeze foods that have been at room temperature for more than a couple of hours; these can be cooked immediately and eaten. If the food has been allowed to thaw at room temperature for longer than 2 hours, it should be discarded without tasting.

How to Freeze Fruits and Vegetables

Although fruits and vegetables are frozen using general techniques, some require particular steps prior to freezing to retain the best keeping quality. Refer to table 4 for specific instructions regarding preparation. Remember, harvest only blemish-free, excellent-quality fruits and vegetables for freezing, because freezing does not kill molds, yeasts, and other microorganisms. For best quality, both eating and keeping, fruits and vegetables should be frozen shortly after harvesting.

After it is cured and rubbed with spices, garlic, and salt, pork belly hangs in the root cellar at Fat Rooster Farm. Once it has been hanging for six weeks, the surface will be brushed off and it will be sliced as pancetta.

PART TWO

Preserving
Meats and Poultry

A Duroc cross piglet sniffs for scraps in a pen at Fat Rooster Farm. The farm raises 8 to 10 pigs per year to sell at market, to their CSA customers, and for home use.

" One time, after we quit plowing, and while my brother was cutting the clover to take home with us to feed the animals, I went into the woods, and there was a big rabbit sleeping. Carefully, I bent over it and grabbed it by the neck. He scratched me up pretty bad, but I wouldn't turn him loose. We tied him up in the coat sleeve, but by the time we'd gathered up the clover and loaded it on the wagon, he'd chewed a hole right through the sleeve and escaped. Oh how wonderful it would have been if we could have eaten rabbit, so rare to us, especially during the War. In the summertime, it was alright, because we had lots of fruit and vegetables to eat, but in the winter it was rough. We had to eat potatoes, beans and sauerkraut with little very little pork and some old hen once in a while. We had to eat whole rye bread so the rye would last until the following harvest. "

– Geza Megyesi, 1909, *translated by Louis Megyesi and edited by the author*

(7)

Raising Animals with Food Preservation in Mind

RAISING YOUR OWN LIVESTOCK for preserving is extremely rewarding. It gives you a sense of responsibility and comfort to know that the meat you've raised for yourself and your family has been well cared for and given quality feed. Raising your own meat can offer the assurance that quality and freshness are guaranteed, and that no corners were cut in the handling of the final product.

Just as you planned your vegetable and fruit variety selections, you'll need to develop a set of criteria to choose which livestock breeds to raise. There are fads that sweep the country, so be aware of being sold the best and the brightest at the moment; Scottish Highlander (Highland) cattle are touted as being cold hardy and great foragers, but they are slow growers and generally not that easygoing to work with if you'd like a steer that the children can have a hand in raising. Cornish-Rock cross meat birds are fast-growing eating machines, but they tend to forage less than Buff-Silvers or Kosher Kings and often keel over from heart attacks or develop breast blisters from sitting on the ground too long. Sheep and pigs grow better when they have compatriots.

Choose livestock that is best suited to both your situation and your pantry needs. Do you really have a setup for a pig, or can you skip the pig and barter in exchange for half a hog with a farmer set up to care for one and avoid a disaster like pigs in the rose garden? Do you like lamb enough to get up in the morning for months to make sure that the ewes birth their young without complications? How about the breeding stock? Are you considering a bull that will need care for 12 months of the year, despite the fact that he's needed only every 13–16 months (if all of your cows are on a pasture-based breeding rotation and will wean their calves at 6 months)? Have you thought about what you will do with animals that are no longer useful for breeding purposes? A 600-pound sow can easily become a liability in an operation where she is not confined in a farrowing crate and can easily suffocate her newborn piglets.

Another important consideration to make before running out and accumulating a menagerie to rival that of Dr. Doolittle's is to research whether zoning permits you to raise farm animals. Some cities will allow a beefer but not a pig, or some laying hens but no roosters. A flock of turkeys gobbling on the lawn may seem bucolic and pleasingly country until they start scratching up the neighbor's flower bed or chase after the housecat.

Some types of livestock are harder to integrate with one another. Sheep pastured with goats don't always work; the sheep may be easy to move from one rotational pasture to the next, but typically the goats are more interested in what you're doing and won't follow the flock. Goats tend to tolerate tethering on a line better than sheep, and their diets are dissimilar as well. Goats tend to browse on twigs and bushes more than the average sheep will.

Some breeds of livestock are bred to result in different end products. For example, cows are bred to produce more milk rather than muscle, and some breeds of sheep are better wool producers than milkers or meat breeds. Use table 6 as a general tool to help you determine which breed will best serve your needs. The breeds are classified as best suited for meat, milk, or draught (working) purposes, based on their temperament, growth of carcass, and gross yields of product. Certainly, an Angus cow can be milked, but she was bred to produce calves that gain weight quickly and distribute muscle evenly for a nicely finished carcass. A Jersey calf

BELOW: Barred Silver meat birds eat a grower mash in the pasture at Fat Rooster Farm. The author chose the birds because they are easier to raise in a pasture than other varieties.

Table 5: **Average Yields for Selected Livestock (lbs)**

	Total Yield	Steaks and Oven Roasts
Beef (16–18 Months of Age, 800 lbs Hanging Weight)	420	172
Pork ** (4–7 Months of Age, 180 lbs Hanging Weight)	200	53
Lamb (4–7 Months of Age, 60 lbs Hanging Weight)	30	8
Goat Kid (7–9 Months of Age, 50 lbs Hanging Weight)	20	6
Turkey (22 weeks, 35 lbs Live Weight)	24.5	n/a
Chicken (9–18 weeks, 6 lbs Live Weight)	4.2	n/a
Duck (18–20 weeks, 5 lbs Live Weight)	3.5	n/a
Goose (22 weeks, 20 lbs Live Weight)	14	n/a

* Fat, bone, and offal (liver, kidney, gizzard, sweetbreads, etc.) should always be considered as a potentially usable part of the carcass
** Each pork side will yield 14 pounds of pork belly that can be smoked and cured or used for salt pork

can produce top-notch meat, but the heifers (female calves) are not generally grown to produce beef. Their milk is known for its high butterfat content, and so it would be considered a waste to slaughter a 16-month-old heifer for meat. Of course there are many more breeds of livestock than those listed here (there are more than 800 breeds of cattle recognized worldwide), but if you're just starting out, keep it simple at first, and work with the animals that are most familiar to your neighboring farmers and veterinarians before

	Pot Roasts	Stew/Ground	Fat/Bone/Offal*
	84	84	80
	(hams) 46	16	57
	10	2	10
	4	1	10
	n/a	n/a	n/a
	n/a	n/a	n/a
	n/a	n/a	n/a
	n/a	n/a	n/a

branching out into exotic livestock that might need more special-ized care.

Special Licenses and Permits

If you are planning to raise your own meat, make sure that you have all of the licenses you need through the state and/or federal inspection facility where you will be taking them to be slaughtered and butchered. There's nothing more frustrating than finding out

Pigs are Different

Most people interested in raising a calf for meat are aware that an Angus or Hereford is more likely to fill out faster and have better marbling within the meat, thereby creating a more tender finished product than a Holstein or Jersey calf would. This is not to say that the latter breeds aren't tasty, and in fact, we cross our cattle with some dairy breeds to produce heifer (female) calves with increased butterfat content in their milk (which translates to higher protein and fat for their future nursing calf) and to produce heifers that are a bit more docile than straight beef breeds tend to be. With pigs, the situation is a bit more diluted.

The pig is one of the first animals known to be domesticated by humans as a food source, about 8,000 years ago. They were descended originally from the wild boar, Sus scrofa, and in North America were introduced in 1493 by Columbus, when he brought over 8 individuals on his second voyage. De Soto and other explorers brought more, and soon, the pig roamed as far as Central America, often reverting back to the wild. After the 1800s, a more serious importation regimen was followed, and important breeds from England, Spain, and Portugal such as Berkshire, Big China, and Irish Grazier predominated.

Pork breeds were historically classified as either lard or bacon types; lard types were fed corn and other grains and were fast growing, short-legged, stocky-bodied, fatty pigs. Bacon breeds were longer and leaner, more muscled out, and slower growing, often fed higher-protein, lower-energy feeds such as forage,

there's a waiting list to get into the slaughterhouse when the family pig is pushing 300 pounds.

Be sure and check with local zoning and ordinances before raising your own livestock to make sure that everything you do is permitted.

When to Slaughter

Again, if you're planning on an off-site butcher to slaughter your animal, make sure you've made plans well in advance. Line up the help to load the animal, the trailer to haul it to the facility, the appointment at the facility. Slaughterhouses are busiest in the fall turnips, rutabagas, and apples. The lard pigs were said to have more flavorful meat and better keeping qualities than their bacon brethren, often obtaining over 3 inches of fat on their carcasses. This was sold predominantly as lard (used in every aspect of baking and cooking and as a mechanical lubricant), but also as fatback and salt pork.

With the advent of World War II, the pork industry in the United States was drastically altered. Lard was diverted to support the war effort as machinery lubricant and to build explosives. Because it was no longer available, consumers switched to other fats, like vegetable oils, for their cooking needs. After the cessation of the war, lard was once again available, but the vegetable oil producers lobbied to protect their market by vilifying lard and touting the health of Crisco and margarine over it as a cooking fat (decades later, the reality of trans fats and their pitfalls for human health were to be discovered). With an overabundance of lard, the pork industry switched to predominately bacon breeds. The remaining lard breeds, the Choctaw, Mulefoot, and Guinea Hog, are in such scarce numbers that they are considered endangered breeds of pigs.

As a farmer who has the opportunity to raise piglets, you will most likely find crossbreeds rather than purebred piglets. Crosses of Berkshires, Yorkshires, and Durocs are most common, with Hampshire, Poland China, Tamworth, and Gloucestershire Old Spot being less common. The most important trait of any piglet you purchase should be its behavior: Healthy piglets should be bright, alert, active, and plump. Avoid piglets that appear dopey, aren't curious about what you're up to, and seem thin.

months, when animals are reaching maturity and the weather begins to chill.

On-farm slaughtering needs careful attention as well. The ancient Book of Hours from the Middle Ages depicted slaughter occurring in the winter when it was cold, with fewer insects, and the meat was able to hang and "cure" without spoiling before it was butchered. It is much more sanitary to slaughter the hogs in early spring or late fall here in Vermont than in July, when the heat can destroy a perfectly good side of pork in a matter of hours.

Again, check local and state regulations to determine that on-farm slaughter is an option. At Fat Rooster Farm, we prefer it,

A piglet naps during the commotion in the livestock barns at the Tunbridge World's Fair in Tunbridge, Vermont.

Chickens and Beyond

Poultry, and in particular chickens, are the most commonly kept livestock on the small farm or homestead. They're relatively easy keepers and inexpensive to purchase. Most breeds of chickens that are widely available at feed stores and mail order sources specify if they are more suited as layers (meaning they'll be more efficient producing eggs rather than meat) or meat birds. Some sources refer to dual-purpose breeds of chickens, which are good at doing both. Granted, these breeds will lay fewer eggs overall and produce meat less quickly than their counterparts specifically bred to do one or the other, but they generally tend to be easier to keep. As an example, the Cornish-Rock cross, the most common meat bird crossbreed available, will average a meaty carcass of about 5 pounds in just over 9 weeks. On the flip side, they're poorer foragers than their dual-purpose cousins, and tend to have some horrendous health issues. Because they are so fast growing, they can develop fatal heart problems and blisters on their breasts and hocks (the back part of their legs), and their meat tends to be a bit bland. A white leghorn hen will hands-down outlay a cuckoo marans, but they are flighty birds that don't do as well in cold, freezing temperatures, due to their finely feathered bodies and floppy, large combs that can easily become frostbitten. The cuckoo marans roosters will weigh 5 pounds at 16 weeks and be just as tender and more flavorful than a Cornish-Rock cross will. Refer to the Resource section for a list of poultry companies that will help you find the right breed for your operation.

As far as poultry goes, don't forget all of the other choices: from turkey to Guinea hen, quail to pheasant, ducks to geese, these are all options for the home poultry operation. These birds tend to have more specialized feeding and housing requirements, so be sure to be prepared before purchasing them.

because the animals are not stressed or frightened by being loaded and carried off to a distant, unfamiliar place. However, if the meat is intended for sale, it is usually required to be federally inspected and will need to be taken to the closest slaughterhouse.

How Much Meat Will I Need for My Family?

Determining the number of livestock you need for your family (which converts into pounds of meat) to live comfortably throughout the year is tricky. You will need to figure out an average number of meals that you imagine you will serve during the year. For example, do you eat chicken once a week? What about unexpected guests, or special occasions? Will you be raising livestock for other people's freezers? The average meat chicken, raised for 10 weeks on a growing ration, will dress whole at between $3\frac{3}{4}$ and $5\frac{1}{2}$ pounds. This should be enough roast chicken for four hearty eaters and enough stock from the carcass, after you've simmered it with water and veggies, for a good soup later down the road.

Mortality should also be factored into the number of livestock that you raise. With large animals such as beef cows, milk cows, pigs, and sheep, the chance of animals sickening and dying or being killed by predators is far less than if you intend to raise hens for laying, or turkeys for meat. I like to factor in at least 10 percent mortality for sheep and goats, and at least 20 percent for poultry. And if you absolutely can't live without bacon, you should always factor in the freak accident and raise enough just in case.

Table 6: **Selected Livestock Breeds and Their Uses**

Cattle	Meat	Dairy	Multi-Purpose*	Working Animal	Wool
Ayshire		*			
Balck Angus	*				
Belted Galloway	*				
Brown Swiss		*	*	*	
Charolais	*				
Devon			*		
Dexter			*		
Dutch Belted		*			
Gelbveih			*		
Guernsey		*			
Hereford	*				
Highland	*				
Holstein		*			
Jersey		*			
Limousin	*			*	
Piedmontese			*		
Randall Lineback		*			
Red Angus	*				
Shorthorn			*	*	
Simmental	*				
Swedish Friesan		*			
Tarantaise			*		
Goats					
Alpine		*			
Anglo-Nubian		*			
Barbari	*				
Boer	*				
La Mancha		*			
Oberhasli		*			
Toggenburg		*			

Cattle	Meat	Dairy	Multi-Purpose*	Working Animal	Wool
Sheep					
Barbados Black Belly	*				
Bluefaced Leicester	*				*
Border Leicester	*				*
Cheviot	*				
Clun Forest			*		
Columbia					*
Coopworth			*		
Cormo					*
Corriedale			*		
Cotswold			*		
Dorper	*				
Dorset	*				
East Friesan		*			
Finnsheep	*				
Hampshire	*				
Hog Island					*
Icelandic			*		
Jacob			*		
Katahdin	*				
Lincoln			*		
Merino			*		
Montadale			*		
Navajo-Churro					*
Polypay			*		
Rambouillet			*		
Romney			*		
Scottish Blackface	*				
Shetland			*		
Southdown	*				
Suffolk	*				
Texel	*				
Tunis	*				

Every two hours for 10–12 hours, the author adds apple wood to smoke bacon at Fat Rooster Farm. The wood is pruned from the farm's 12 apple trees.

> **"**It is not necessary that the smokehouse should be very tight, but it is important that the pork should not be very close to the fire. A smothered fire made of small billets of wood or chips (hickory preferred), or of corn cobs, should be made up three times a day till the middle of March or first of April, when the joint pieces should be taken down and packed in hickory or other green-wood ashes, as in salt, where they will remain all the summer without damage of bugs interfering with them.**"**
>
> —**House Keeping in Old Virginia,** *Marion Cabell Tyree, 1879*

⑧

Preserving Meat

TONIGHT I AM CONSUMED *with a task that has largely been lost in our culture—the art of making bacon. It's just pork, just flesh from the belly of the pig, but somehow, it's transformed into a delicacy of indescribable sensation. Do we know how bacon happens anymore?*

Hams are just roasts unless they're cured, and bacon is just pork belly unless it's tended to. Bacon and ham are what they are because of the process they go through—brining, then drying, then smoking, to achieve that achingly buttery goodness. Yes, you can cure bacon and hams without nitrates.

Because I hate what is put into the process of curing a modern-day pork belly, I do it all at home now. The pig is killed here, cured here, and last night, thanks to my husband Kyle, the meat began the last process of being smoked here. He made a smoker out of an old refrigerator. We were lucky, because this was a fridge from the 1960s, metal, ceramic interior, and smokehouse-worthy, so it was well suited for conversion. There was no plastic interior to worry about; the Freon gas hidden inside its interior had been safely discarded.

In order to finish the bacons in the new smoker after they had been properly cured in salt, maple syrup, and spices, I resolve to stay up all night. Bradford has the beginnings of what looks like the swine flu (fitting since I am processing pork), so I have two excuses to stay vigilant tonight. Every 2 hours, I go out to the new smoker, feed the tinderbox with a mixture of dry and wet hardwoods, and fill the reservoir with water. Every 2 hours, I gauge the fever raging in my son, and swab him with wet washcloths and prompt him to drink water.

At 3:36 a.m., after the bacons have turned a golden brown, and I know that I will be awake in just 2 more hours, and Bradford's forehead is cool with the sweat of a broken fever, I crawl into bed with two cats and Kyle. It is warm and cozy,.

At 4:13 a.m., this all abruptly ends, with Kyle's shouts of "It's on fire, the smoker is on fire!"

I open my eyes to see flames reaching up to our bedroom window, almost 15 feet high. I run out, with a bowl of water, for a grease fire, and throw it on the inferno, only to make everything worse. Kyle appears with the fire extinguisher and gives a mighty blast, then another and another.

In my head, I am still thinking that we can save the bacons. I'm yelling for him to stop with the chemicals, that I can put out this inferno, which is now melting the ceramic, with water. It's roaring. Reaching up and over to the bush that lies next to the propane tank that feeds our cooking stove. "They're gone, they're gone, it's gone," he says, with another blast. He fought forest fires for years . . . I am out of my element. I sulk back to the house and try to sleep for 2 hours before my day has to begin. I don't know what has gone wrong. Did the grease from the bacons cause the hot plate that was heating the green woodchips to flame up? Did the string used to tether the bacons to the hanging wire fail and topple the bacon onto the hotplate?

Kyle joins me in bed after he is sure that the fire has been put out and begins giggling at his poetry:

Autumn, and the smell of apple wood—smoked bacon drifting toward the bedroom window.

The sun is up, flowing in from the south.

Wait a minute—it's four o'clock in the morning, it's too early for the sun to be up, and from the south???

Be quiet, I say.

Bacon.

A try at running a homemade smoker didn't go well at Fat Rooster Farm. A hot plate in the old refrigerator caught fire, burning up the meat and everything else.

The History of Smoking and Curing Meat

Curing and smoking are some of the oldest forms of food preservation. The key ingredient in a cure, salt, was perhaps used first by the Egyptians to salt and preserve their dead, turning them into mummies. They also used salt to soak the inedible olive, turning it into a Mediterranean mainstay. Fish, notably tuna and herring, were cured by salting and then smoking as long ago as 2000 B.C.E. The Celts and, in turn, the Romans brought cured pork into France and Germany near 1000 B.C.E. Charcuterie, the French art of sausage making, relies on brines, dry cures, and smoking to create some of the most flavorful preserved meats in the world; Germany boasts well over 1,000 types of sausages (like Mettwurst and Teewurst), as well as the more familiar bologna and frankfurter. Probably no sausage eater has been without sampling *peperone* (pepperoni), originally from Southern Italy,

In colonial North America, meat was only slaughtered in warm months when it would be eaten right away. Winter months traditionally marked when meat, typically pork, was salted and then

How does Salt Cure Meat?

Attracted by electrically charged sodium ions, water rushes out of the semipermeable cell walls of the meat to equalize the concentration of liquid. In the process, the cells are depleted of water (and dehydrated), and the concentration outside the cell fills with blood and water from the cell. Proteins remain within the cells, because they are unable to pass through the cell wall, but their structure is changed because of the ionic change occurring during the back-and-forth shifting of ions. Microbes within the cell are also dehydrated as the water is pulled from their cells and are killed or rendered incapable of multiplying. Think of it like sucking on a lemon. The tartness of the lemon juice, higher in concentration than saliva, makes you pucker and your mouth begins to water, equalizing the concentration of liquid within your mouth.

cured in the "Smoakhouse." Much like today, where the majority of households own a refrigerator, early Americans had smokehouses in the back yard that were used to smoke the meat and then keep it safe from insects and vermin. Whole hams could be stored throughout the year, and many of them hung to cure for at least two years before consumption. The traditional Virginia hams had to be soaked in several changes of fresh water before eating, due to the high salt content of the cure.

Bacteria responsible for spoiling meat and causing food-borne illnesses such as salmonellosis and botulism thrive in meats that are kept at temperatures above 40 degrees Fahrenheit (see page 223). The temperature danger zone for bacterial growth is between 50 and 130 degrees Fahrenheit; meats kept at temperatures warmer than 70 degrees Fahrenheit (the average temperature the counter shelf tends to stay at during most of the year) can lead to bacteria multiplying 700 times within 12 to 24 hours. At 80 degrees Fahrenheit, they're capable of multiplying 3,000 times within the same time period.

The moisture content of meat can also add to the rate of spoilage. Water acts as a medium in which bacteria can grow and move throughout the tissue. Curing acts to preserve meat by drawing water out and replacing it with salt. Smoking coats the meat with a tarry film called creosote that act as a seal on the meat's surface to protect it from infiltration by bacteria, insects, and molds.

Curing

Meats can be cured and preserved using either a dry rub or brine. Dry rubs can be a combination of salt, sugar, and spices, rubbed over all of the meat's surface. The meat is then kept for a period of time in a cool (below 40 degrees Fahrenheit), nonreactive container. Colonial Americans used truncated wooden barrels with several large, drilled holes in the bottom. The moisture drawn out of the meat by the salt would drain through the holes.

Brining, also called pickling or corning, is an easier method to cure meat. Brines have the advantage over dry cures of penetrating deep within the meat as well as distributing the flavors throughout the meat. Brines can be a mixture of sugars, salts, curing salts like sodium nitrate or potassium nitrite, and spices. Curing salts containing nitrates are only necessary in brines when the meat will be exposed to temperatures above those typical of a refrigerator for more than 1 or 2 hours, so if you aren't smoking or air-drying the meats for extended periods of time, they're not necessary.

One common combination is a 10 percent brine solution made by combining 1 to 1½ cups of kosher or pickling salt to 1 gallon of liquid. In historic times, a general rule of thumb to be sure that the brine was sufficiently strong enough was to float an egg or potato in it. If the object stayed just afloat, the brine was good.

· · · · · · · · · · · For Curing Ham · · · · · · · · · · ·

For five hundred pounds of hams.

- *1 peck and 1½ gallons fine Liverpool salt*
- *1¾ pounds saltpetre*
- *1 quart hickory ashes well sifted*
- *1 quart molasses*
- *2 teacups cayenne pepper*
- *1 teacup black pepper*

Mix these ingredients well together in a large tub, rub it into each ham with a brick, or something rough to get it in well. Pack in a tight, clean tub and weigh down. Let hams remain six weeks; then take them out and rub each one on the fleshy side with one tablespoonful black pepper to avoid skippers. Hang in the meat house, and smoke with green hickory for from ten to twelve hours a day for six weeks, not suffering the wood to blaze. On the 1st of April, take them down and pack in any coal ashes or pine ashes well slaked. Strong ashes will rot into the meat.

—Mrs. R.M. in **House Keeping In Old Virginia,**
by Marion Cabell Tyree, 1879

It is easy to make foods too salty by brining them too long. Follow the brining instructions for the specific recipe to the letter to avoid failure. A simple way to test whether the meat is too salty is to slice off a piece and fry it. Of course, being on the outside, the meat will taste a little too salty, but it shouldn't be inedible. Over-salted meat can be soaked in unsalted water for a period of time—usually about half the amount of time that it was in the brine—or you can change the way you cook the meat. A brined pork roast can turn into a braised roast, surrounded with unsalted cooking liquid and root veggies that will soak up the excess salt.

After brining, the food should be allowed to sit to soak up the migrating solution throughout. For a large, whole ham, this may mean resting in cold storage (above freezing, but less than 40 degrees Fahrenheit) for up to 24 hours, while boneless chicken breasts will require just a couple of hours to sit.

Never reuse brine. The solution will have been changed from the original as far as salt to water ratio, and it will be contaminated with bits of fat and muscle that could cause the next piece of meat being prepared within it to go rancid.

Curing Equipment

Be sure to use nonreactive containers when brining or preparing foods that have been dry cured. Food-grade plastic buckets make excellent brining containers, as do food-grade tubs. You need to remember that the container must fit into a refrigerator or walk-in cooler or root cellar that is kept below 40 degrees Fahrenheit.

Use pickling or kosher salt for curing; never use table salt. Select fresh ground or fresh whole spices. You can easily prepare your own herbs from ones you've previously dried by pulverizing them in an herb mill or mortar and pestle. I also prefer using maple syrup to granulated sugar in my brines; some recipes prefer dextrose.

Smoking

Before foods are smoked, they should be allowed to dry. Brined foods should be patted dry and left in the refrigerator or cool place (below 40 degrees Fahrenheit) so that they become tacky, or form what's referred to as a pellicle. Foods that aren't properly dried will still take on a smoky flavor, but the tacky surface allows the particles in the smoke to firmly adhere to the food being smoked.

Food can be preserved using either a cold-smoking or a hot-smoking technique. As the name implies, cold-smoking doesn't actually use the heat of the fuel to cook the meat; instead, the steady drying of the cold-smoking process preserves the food. Cold-smoking is much more difficult to achieve than hot-smoking; temperatures that do not exceed 100 degrees Fahrenheit and are more typically maintained between 80 and 90 degrees Fahrenheit are used in this process. The amount of time food is cold-smoked can range from 24 hours to 3 weeks, depending on the size of the food and the steadiness of the smoking temperatures. Unless you have a smoking apparatus that can reliably maintain temperatures below 100 degrees Fahrenheit, I wouldn't recommend cold-smoking as one of your first projects.

BELOW: The author cuts pieces of green apple wood to add to two smokers at her farm. The wood is pruned from the farm's twelve apple trees.

After brining or dry-curing and cold-smoking, the food is hung in a cool, dry place (about 60 degrees Fahrenheit) for a period of between a few days to a few weeks before it is stored or consumed. Cold-smoked foods can remain fresh up to 2 weeks in the refrigerator or they can be frozen immediately.

Hot-smoking refers to those foods that are fully cooked at temperatures at or above 150 degrees Fahrenheit. Unlike grilling, hot-smoking is maintained at lower temperatures over a longer period of time, and unlike cold-smoked foods, the end product is not dry enough to safely store it without freezing for long-term preservation. Guidelines for hot-smoking temperatures vary widely; some recipes recommend maintaining temperatures at about 200 degrees Fahrenheit for whole cuts, while others prefer temperatures as high as 300 degrees Fahrenheit. If your setup is a backyard grill with an instant read thermometer, you may want to concentrate on achieving a nice, golden smoked color and then finish the food off in the oven until it is cooked through to an internal temperature that is appropriate for the cut.

Home Cured Meat Safety: Salmonella, Campylobacter, Clostridium, and Staphylococcus

Many homemakers have been scared away from home curing and sausage making because of the potential for food-borne illnesses. All five of the most common bacteria responsible for "food poisoning" occur naturally in our environment as well as in our foods. The danger occurs when conditions are present that cause these bacteria to multiply and grow to the point that they'll cause sickness, even death.

Salmonella is by far the most common source of food-borne illness and is usually associated with improperly handled poultry products. As long as food is heated above 140 degrees Fahrenheit for 10 minutes, the bacteria cannot survive. Keep meat and vegetable products impeccably fresh and chilled below 40 degrees Fahrenheit; only consume fresh, properly prepared foods to avoid salmonellosis.

Staphylococcus aureaus and Clostridium perfringens can both affect foods that have been held at improper temperatures (above 40 degrees Fahrenheit and below 140 degrees Fahrenheit).

Campylobacter infection is the most common cause of diarrhea-related illnesses in the United States. It, too occurs as a result of handling raw or undercooked poultry products. Freezing will reduce the number of bacteria on the meat, and cooking breast meat to 170 degrees Fahrenheit and dark meat to 180 degreees Fahrenheit destroys the bacteria.

Clostridium botulinum causes perhaps the scariest of all food borne illnesses, and is named after the word botulari, which is Latin for sausage maker or seller. The bacterium earned its name after being traced back to fermented sausages containing the toxin produced by *C. botulinum*, which killed men in the Roman legions after they had consumed the tainted sausages on the battlefield. The toxin is so deadly that even a taste on the tip of your tongue from tainted food is enough to cause death.

Fortunately, it is extremely rare; fewer than ten people are sickened by botulism in the United States annually (thousands of people

BELOW: Pork belly is smoked with apple wood to be turned into bacon at Fat Rooster Farm. Most bacon sold in the supermarket is cured, but not smoked, according to the author.

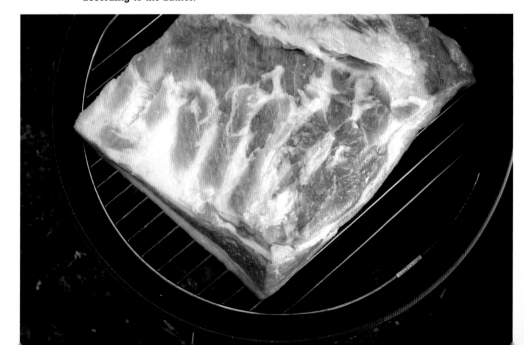

are sickened, and even die, due to outbreaks of *Escherichia coli* in everything from spinach to peanut butter to fresh salsa). Most cases occur when home-canned peas and beans or fish that have been improperly preserved are consumed, and more than 13 percent of the cases occur in Alaska, where meats have been improperly handled and prepared. Deaths from botulism have decreased with the advent of respirators and other life-support advances.

The spores of *C. botulinum* are not what cause the illness; rather it's the toxin the bacteria produce that is deadly. Maintaining the food at a temperature above 176 degrees Fahrenheit for 10 to 20 minutes after opening will render the toxin inactive (it will not kill the bacterial spores, though, which could potentially create more toxin if the food is contaminated and left to consume for later without reheating).

It is for this reason that nitrates and nitrites, or curing salts, are now added to commercially cured meats—everything from hot dogs to bologna have these food preservatives. Historically, meats were brined or dry-cured with just salt. Pure salt (sodium chloride) turned meat gray and rock salt would cause the meat to turn tough. By experimentation, some rock salts with impurities were discovered to have the ability to keep the meat moist and retain its vibrant pink color. These impure salts contained potassium nitrate, known as saltpetre or saltpeter. Saltpeter was widely used as part of the "three S curing," salt, saltpeter and smoke, as long ago as the Middle Ages, but saltpeter's inconsistent results were deemed too unpredictable in the United States by the 1970s, when sodium nitrates (and nitrites) came into favor (in Europe, potassium nitrate is still widely used).

Nitrates and nitrites are controversial because they will sometimes create compounds called nitrosamines, which have been suspected as cancer-causing agents. One need only to search the Internet for literature on these food additives to see how highly charged a controversy nitrates and nitrites remain.

For meats cured and smoked in the home, as long as they are not exposed to temperatures above 40 degrees Fahrenheit longer than 1 or 2 hours, nitrates and nitrites are not needed. Salting, high pH conditions, and competing microorganisms like lactic-acid—producing bacteria can all prevent bacterial growth, even in *C.botulinum.*

Smoking Equipment

Depending on how serious you want to be about the craft, you can use anything from a Weber grill to a stone smokehouse to smoke foods. Smoking can become addictive, no pun intended, just because the end product that you can achieve is far superior to the "maple-smoked" bacon available at the grocery store, for example. However, if you're not into fussing and fretting about the amount of smoke and the temperature inside the smoker, you may want to

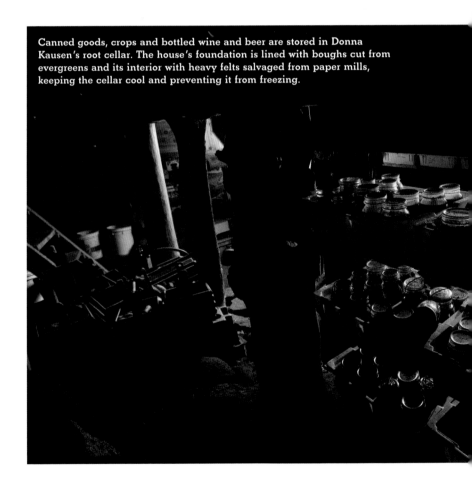

Canned goods, crops and bottled wine and beer are stored in Donna Kausen's root cellar. The house's foundation is lined with boughs cut from evergreens and its interior with heavy felts salvaged from paper mills, keeping the cellar cool and preventing it from freezing.

preserve your foods using a different method. Smoking requires constant attention and detail.

How serious you are about smoking foods will determine what type of smoker you need. They can range from rustically prepared steel barrels fitted with screens and a hole in the base leading to a fire box to a stainless steel, automated smokebox costing several thousand dollars. The Resources section lists several resources for smoking equipment. In rural areas, almost all hardware stores carry inexpensive smokers that are generally reliable for hot-smoking. You might want to start with one of these to see if the craft is something you enjoy before going out and spending lots of money or constructing an elaborate outbuilding that will end up as a toolshed after you've lost interest.

Canning Poultry and Red Meat

With the advent of refrigeration, canning meat has fallen out of favor in most households, yet if done properly, it will yield a high-quality food that contains far less sodium than cured and smoked foods, and the need for keeping a freezer running year-round is eliminated. To be sure, freezing meat is more time efficient, and modern freezing techniques can keep meat safely for over a year.

All meats should be processed in a pressure canner at 240 degrees Fahrenheit to destroy any harmful bacteria.

If you choose to can meat, it is best done using animals that you have raised and slaughtered or that you purchase directly from the farmer, so freshness and quality can be assured (many meats in the grocery store can be several days old). Most meats should never be brined or soaked before canning, and beef, lamb, and venison should not be aged for more than 48 hours prior to canning.

Liquid is never added to the jars of raw-packed meats, nor are thickeners of any kind, including gravies. Meats that are to be used in other recipes should have few added spices and no vegetables (the amount of time required to can meat safely will ruin the nutritional value of the vegetables and render them tasteless and soft).

Poultry and Rabbit—Raw-Pack—Process other small game such as squirrel or wild gamebirds using these methods. The depth of flavor and tenderness of old roosters or laying hens that have been canned is perfect for stews and soups.

Allow poultry to chill below 40 degrees Fahrenheit for between 6 and 12 hours; rabbits and squirrels should be soaked 1 hour in salted water (4 tablespoons per gallon) and then rinsed.

Prepare the jars and then cut the bone-in pieces to fit the jar. Trim off excess fat.

For poultry, cut the meat at the joints; debone the breast, but leave the bone in the wings, thighs, and legs. Loosely pack the jars with the deboned breasts in the center of the jar and the thighs and legs on the outside of the jar. Add up to 1 teaspoon of salt per quart. Allow 1¼ inch headspace and process in a pressure canner for 1¼ hours for quarts

Ground Meats and Sausage—Hot-Pack—Unless you're planning to grind the meat yourself, don't can ground meats; for lean meats such as moose, veal, or venison, add 1 part pork fat to 3 or 4 parts lean meat before grinding.

Cut cased sausage in to 3-inch sections or shape ground meat into patties or balls. Cook in butter or olive oil until lightly browned (the meat can also be browned without shaping, but it's harder to remove from the jar). Pour off all the fat, or it could give the food an off flavor.

Fill the jars with pieces and add boiling meat broth, tomato juice, or water, allowing 1 inch headspace. Add up to 2 teaspoons of salt per quart (less if using tomato juice, since that's already been salted) and process in a pressure canner for $1/4$ hours for pints and $1\frac{1}{2}$ hours for quarts.

Chunks and Strips of Meat—Raw Pack—Use only fresh, high quality, chilled meats, with all excess fat removed (the fat can impart a strong off flavor to meat that has been canned). It is acceptable to soak meat from wild game that is strongly flavored (like bear, boar, or venison meat from a buck) for up to 1 hour in brine made from mixing 4 tablespoons of salt per gallon of water. Rinse the meat before proceeding.

Fill prepared jars with raw meat, leaving 1 inch headspace; add up to 2 teaspoons of salt per quart to each jar, but don't add liquid.

Process in a pressure canner for $1\frac{1}{4}$ hours for pints and $1\frac{1}{2}$ hours for quarts.

Drying Fish and Meat

Since prehistoric times, the art of drying fish and meat was relied upon to sustain communities through the winter bare months as well as supplying voyagers and other travelers on their journeys. More than 2,000 years ago, the Chinese regarded dried snake as a convenient dried "fast food," while Europeans seriously exploited the resources of Newfoundland for their cod that was dried, salted, and transported across the seas as early as 1497. In the Americas, two dried meats were especially important. Pemmican, derived from the Cree Indian word for fat, was especially suitable for the cold northern climate. It consisted of the thinly sliced dried meat of large game animals, which was dried using fire, sun, or wind. The meat was then shredded by pounding it and then mixed with bone marrow, melted fat, and fruit like red currants or wild cherries. The mixture was packed into rawhide containers, sewn shut, and coated with more melted fat.

Charqui, originating in Peru, was made by preparing strips of beef that had been dipped in brine or dry-salted and then wrapped in animal skins for about 12 hours, so that the meat released some of its juices. It was then hung in the sun to cure and stored bundled until use. The strips were usually pounded between stones to soften them before eating. Our modern word for jerky is derived from this word and ancient method of preservation.

There are several other regional examples of dried and cured meats: *biltong* from South Africa, *bindenfleisch* from Switzerland, *bresaola* from Italy. All are examples of how lightweight, preserved meats could be carried along for months or miles at a time to sustain a varied diet of protein.

Brining and dry curing as well as smoking are usually methods employed to dry meats. Oily cuts of meat tend to require more elaborate methods to dry, while lean meats can commonly be dried using just the sun and wind. Modern curing techniques now typically include salt, spices, and herbs, and some form of potassium nitrate (see page 220).

Simple rules of thumb to follow in order to dry fish and meat properly are 1) use only fresh, impeccably clean meats that have been kept at temperatures lower than 40 degrees Fahrenheit during the aging process; 2) prefreeze meats that have dry-cured and are not fully cooked before eating to prevent ingestion of parasites. The freezing will desiccate, or dry out, and kill any parasite eggs in the meat. Last, home curing should follow the recipe to the letter; experimenting with a roasted root vegetable dish by altering an existing recipe can be a flop, but skipping a step in making salami can prove fatal.

The Resources section lists several good resources for the homesteader who becomes serious about charcuterie, or the craft of drying and curing meats. If you don't feel so adventurous, freezing meats is hands-down the easiest way to preserve fish and meats.

Freezing Fish and Meat and Seafood

For general freezing instructions, including methods, packaging, storing, thawing and refreezing, refer to Chapter 6.

Freezing fish, meats, and seafood offers the most nutritious, safest, healthiest, and most economical means of preservation for these foods. Canning causes a loss of nutrients and is not ideal for tender cuts of meat and seafood; smoking and curing are controversial because of the high sodium content and additives needed to preserve these foods effectively. Whole carcasses or sides of meat will easily fit in a chest or upright freezer, allowing an economy of scale that isn't possible using other preservation techniques.

Just as for fruits and vegetables, freeze only the highest quality fish, meat, and seafood. The packages should be frozen in meal-sized portions. For fish and seafood, freeze in 1- or 2-pound packages; for meat cuts, freeze in portions to feed two to four people. Make sure that excess fat has been trimmed from individual cuts of meat; fat doesn't keep long in the freezer before taking on an off, rancid flavor. For poultry and other fowl, freeze whole or cut up, but don't stuff or freeze the innards inside the cavity. Make sure that poultry and other fowl has chilled to 40 degrees Fahrenheit before freezing or it will become tough. Fish that is less than 2 or 3 pounds before it's dressed should be frozen whole; fillets retain their best quality and texture if sliced no thinner than ¼ inch thick. Hard clams, oysters, and scallops can be frozen live in the shell or shucked, but steam open any softshell clams and mussels, then freeze. Crabs and lobsters should be boiled and cooked, then the meaty pieces can be frozen whole in the shell or picked and frozen. Shrimp are best if they're frozen raw with their heads removed but shells still intact.

Any meats eaten within a couple of weeks can be packaged safely in freezer bags or rigid plastic containers. Any longer storage, and the foods should be double wrapped with wax-coated freezer paper and secured with freezer tape (don't use other types of tape or it will become unstuck in the freezer).

Most frozen fish and meats can be cooked thawed or unthawed, though cooking time for unthawed meats may take as much as 1½ times as long for doneness. Don't cook foods that will be stuffed, breaded, or coated, any seafood except shrimp (unless it will be breaded), or crayfish until it's completely thawed. Thaw these foods in the refrigerator or in a location that is 40 degrees Fahrenheit or lower. Don't let them thaw above other cooked foods that will be dripped on; wrap them in plastic and keep them in a container to catch any thawed fluids that result.

Pickled eggs in the pantry at Fat Rooster Farm. The eggs are from Nancy, a modern game bantam hen kept in a cage in the farmhouse living room.

9

Preserving Dairy Products and Eggs

CULTURING FLUID MILK HAS been practiced since at least 5000 B.C.E., as evidenced by cave paintings depicting nomadic herdsmen making cheese. It is not precisely known who first discovered the keeping qualities of milk left to ferment and separate into curds—perhaps it was a herdsman on the Sahara transporting his milk in a pouch made from the stomach of a calf (rennet occurs naturally in the stomach of ruminants and is responsible for curdling milk) or maybe it was just a pail of milk left out too long in the heat of the day that prompted some adventurous soul to taste it and find that it was sumptuous.

Regardless, it is more likely to find traditional foods made from cultured and fermented milk products than it is to see it used in its natural, unfermented state. This stands to reason, since before the age of pasteurization and refrigeration, fresh milk would quickly sour and separate due to the production of lactic acid from the

Lactose Intolerance

During fermentation, as much as 40 percent of the lactose in milk is broken down into lactic acid. In unpasteurized milk, the presence of the enzyme lactase largely aids in the digestion and decomposition of casein, a protein that some individuals find difficult to digest. If you find yourself unable to tolerate fluid milk, try live cultures of yogurt and other dairy products, such as kefir and whey.

milk's sugars and proteins. Historically, different lacto-fermenting bacteria were responsible for this process, and the end results were not always predictable. Now, milk is first heated to kill off all of the bacteria present and then inoculated with a culture specific to the desired result. Some cheeses, like the Swiss cheese Gruyere, demand controlled environments and inoculation with several types of bacteria to achieve their meltingly buttery taste. Still, culturing dairy products at home like butter, cream cheese, cottage cheese, mozzarella, yogurt, and buttermilk is entirely feasible and fun.

Making Butter and Crème Fraîche

As a junior in high school, I made butter in a quart jar in front of a class of very bored English students. No one was particularly impressed, given that going to the grocery store and choosing from a dairy case full of products involved much less effort. Homemade butter is much more tangy than its store-bought equivalent. And sharing how it's made with a class of third-graders is much more fun.

In order to make butter, you must have cream. Butter made from cream that has been collected during the months when cows are grazing on fresh, lush vegetation is by far superior to that made during winter months or from pasteurized cream. For best results, seek out a local supply of raw, whole milk and skim off its cream. Generally speaking, one gallon of raw, whole milk will yield about one quart of cream and ultimately about ½ pound of butter. Cream can be aged so that lacto-fermenting bacteria have time to break down and ripen it, or it can be churned fresh to create sweet butter. I prefer the taste of butter made from cream that has been allowed to sit at room temperature for about half a day.

Making butter with a butter churn is probably the most hassle-free method to use; the paddles are either wooden or plastic, and the top has a perforated area from which the leftover liquid (the true buttermilk) can be poured off. A food processor with a metal blade can also be used, as can a glass quart jar with a well-fitting lid.

By agitating the cream, the fat globules within the cream are forced together into a solid, squeezing out the water, dissolved milk sugar, and proteins. Whey differs from buttermilk because it is the end product from milk (which may or may not contain cream) rather than the end product from just the cream.

After agitating or churning the cream and the butter has formed, strain off the liquid buttermilk and put the butter in a bowl. Use a flat spoon or butter paddle to force the rest of the buttermilk out—the more you get out of the butter, the longer the butter will keep. The butter can now be transferred to a mold or butter crock and kept cold. Unlike whole milk, butter freezes well.

Another easy trick is to make crème fraîche, the gourmet sour cream used in many French cooking recipes. You'll need cultured buttermilk (not the liquid from the butter you've made) and 2 cups of cream, preferably raw and unpasteurized. Add 1–2 tablespoons of cultured buttermilk to the cream, stir it well, and put it in a glass jar. Cover the jar tightly, and put it in a warm area for a day (near the woodstove or in the oven with the pilot light on). After aging, keep it chilled by storing it in the refrigerator or a root cellar kept cooler than 40 degrees Fahrenheit.

Making Cultured Buttermilk and Yogurt

It's next to impossible to find cultured buttermilk offered for sale in most grocery chains nowadays, and the list of ingredients in most of the "yogurt" reads like a horror novel. Go-GURT brand "yogurt" made by Yoplait may just have taken the marketing deception prize by supplying a seemingly healthy snack for parents to feed their children that is actually higher in sugar per ounce than Coca-Cola.

· · · · · · Is Raw (Unpasteurized) · · · · · ·
Milk Safe?

The raw milk controversy is one of the most loaded food safety questions there is. Advocates for raw milk cite its ability to offer a "protective effect" against some allergic conditions like asthma, hay fever, and eczema if consumed during childhood. This makes some degree of sense, since the more exposure to pathogens we're given, the greater our immune response becomes. Nutrients are lost during pasteurization of milk products, so raw milk does contain more beneficial enzymes, vitamins, and minerals than its store-bought cousin.

Those in the "pro" raw milk camp attribute its fall from popularity to a series of articles published in *Ladies' Home Journal, Reader's Digest,* and other magazines during the early 1940s. Undulant fever (brucellosis) was said to have wiped out scores of people in a town called Crossroads, which was later found to be fictitious: no outbreak actually occurred.

When the practice of feeding dairy cattle spent grains from distilleries began, undulant fever outbreaks were more common and were responsible for the deaths of many children and immuno-compromised adults. Instead of grazing on lush pasture with exposure to sunlight and fresh air, these cows were confined to barns near the distilleries that were dark and manure-filled—a breeding ground for disease. Feeding cattle grain changes the pH in the stomach and causes abrasion in the intestinal tract. The intestinal lining

Both buttermilk and yogurt are simple to make at home. You can use sheep's, goat's, or cow's milk, skimmed or not, pasteurized or raw (don't use ultrapasteurized or homogenized products).

The simplest way to make buttermilk is to start with cultured buttermilk that you've already made or from the store. Add about ¼ cup for every quart jar that has been cleaned and sterilized, then becomes thinner, allowing more bacteria and parasites to enter the bloodstream and milk.

The overwhelming "cons" of consuming raw milk products are food safety-related hazards. Along with brucellosis (which has nearly been eradicated from most of the developing countries), raw milk can harbor bacteria from infected teats, sloppy sanitation, and contamination from manure and human carriers. These can cause foodborne illnesses like campylobacter, salmonellosis, E. coli O157:H7 infection, and listeriosis, to name the most common.

The bottom line: Consult your physician if you are elderly or immuno-compromised. Talk to your pediatrician and ask for hard evidence for or against the use of raw milk.

Know your cows, know your milk. Not all raw milk is equal. Certainly, confinement dairies with thousands of cows that are given growth hormones, antibiotics, dewormers, and milk production enhancers should not be your source of raw milk. These animals are in an environment where bacteria can abound, and only by pasteurizing the final product can you be certain that it's safe to drink (it's kind of disturbing to realize that dairies are allowed to have a certain percentage of bacteria and "pus" in their product, provided it's to be pasteurized).

Talk to your local farmer. Ask her or him if the milk could be tested by the same regulatory offices that test the larger dairies. Look at the conditions that the cows are kept in. What are they fed, and how are they cared for?

Dairy products are no different from other foods: they should be impeccably fresh and handled carefully and cleanly. Take control of your food source; don't let others do it for you without your permission!

top off with pasteurized or raw milk (if using raw milk, be sure that it is fresh and has been handled cleanly from udder to pail). Seal the jars tightly with clean caps, shake the contents, and put the jar in a warm place (around 70–75 degrees Fahrenheit) for 24 hours. The milk should be transformed into a thicker, buttery liquid and should taste tart and sour but not spoiled. Store the buttermilk in the refrigerator, where it will keep fine for up to 2 weeks. Add a little salt to the buttermilk if you're drinking it plain.

To make yogurt, you can use a prepared culture, start with commercial live-cultured yogurt (not Go-gurt!), or yogurt from one of your previous batches. The more starter you use to make the new

Pickled Eggs

These are an easy stand-in for deviled eggs, which I find harder to make and messier to take to a picnic. They'll keep for up to 3 months in the refrigerator, and you can easily seal the jars for keeping in the pantry for up to a year.

Place in each wide mouthed quart glass canning jar:
- Hard-boiled eggs (8-12, depending upon size, leaving ½ inch headspace)
- 2 whole, peeled cloves garlic
- 2 whole red peppers
- I teaspoon mixed dried herbs, like basil, dill, rosemary, or oregano
- I teaspoon black peppercorns

Pour into sterilized jars to ½ inch headspace the following mixture that has been heated to near boiling:
- 4 cups white vinegar
- 3 cups water (nonchlorinated)
- I teaspoon mild curry powder
- ½ cup kosher salt

Seal the hot jars with metal lids and rings and place them in a boiling water bath. Process the jars for 15 minutes; remove and let them cool. Check the seals to see that they're closed; unsealed jars can safely be kept in the refrigerator.

Let the eggs steep in the brine for at least 3 weeks before eating. Makes 4 quarts.

Variation: Use small brussels sprouts or snap beans that have been rinsed and trimmed. Process the jars for 25 minutes to soften but not fully cook the sprouts or beans. Let age at least 3 weeks before eating.

batch of yogurt, the thinner and more sour the end product will be. Depending upon how you want to use the yogurt—as an accompaniment to fatty meats, or as part of breakfast—you'll want to experiment with the amount of starter you use.

You can buy fancy yogurt makers with temperature controls, but it's just as easy to start simply and use what you have at home.

The difference between buttermilk and yogurt is the temperature at which the milk is kept during its transformation. In order to encourage yogurt to thicken, a bacterium called *Streptococcus thermophilus* helps out the other lactobacilli, and it needs temperatures above 110 degrees Fahrenheit and ideally between 110 and 120 degrees Fahrenheit in order to proliferate.

Start with 1 quart of milk that has not been ultrapasteurized or homogenized, preferably raw milk from a clean, reliable local dairy. Slowly heat the milk from 120 to 180 degrees Fahrenheit, then cool it to a temperature between 90 and 110 degrees Fahrenheit. Transfer the cooled milk to a clean, sterilized glass jar (wide mouth canning jars work well), and stir in your starter—about 1 tablespoon for milder, creamier tasting yogurt, and up to ¼ to ½ cup starter for a thinner yogurt with a kick. Seal the jar tightly with a lid and then let it sit for 8 to 12 hours in a warm place, ideally 110 to 115 degrees Fahrenheit. The longer the yogurt sits without refrigeration, the more fermented and sour it will become. When stored in the refrigerator, it will keep well for 10 days to 2 weeks before turning more sour tasting (this isn't harmful, it just may not taste as good with fresh strawberries for breakfast).

Preserving Eggs

Eggs are easily preserved by pickling, freezing, and drying. It is far easier to pickle eggs and can them or freeze them raw. However, powdered eggs come in handy if you've run out of storage space or need your food to be lightweight.

There are several opinions about how best to preserve eggs that have been dried. Some say that the potential for salmonella poisoning is too great to risk drying raw eggs. They advocate first scrambling the eggs and then proceeding with drying them. Others suggest separating the whites and the yolks and drying at 105 to 115 degrees Fahrenheit, or drying the raw, scrambled eggs at the highest setting. All attempts to dry eggs should be made with a dehydrator that is equipped with a thermostat to ensure the safest final product.

For cooked, dry eggs, first scramble and then cook the eggs in a skillet with a little butter or olive oil to keep them from sticking. After cooking, dry them in a dehydrator until brittle. Pulverize the eggs in an herb grinder or food processor until fine. Keep them in the freezer for long-term storage; they will keep fresh for camping and hiking for a week if stored in an airtight container.

To dry raw eggs, I suggest separating the thinner whites from the yolks, and then recombining the two after they've dried. The eggs should be very fresh, with firm, sticky whites and rounded, yellow yolks that remain whole when separated. Beat the egg whites until they form a stiff meringue, then proceed with drying them in a dehydrator at about 115 degrees Fahrenheit for about 10 hours. They should be brittle and crisp. For the yolks, beat until frothy and thick. Dry them in the dehydrator at the same temperature. After they've become brittle and crisp, combine the two and grind until powdered in a food processor. A coffee grinder that is designated for grinding herbs is ideal for this.

For freezing, eggs need to be shelled and frozen raw. Ice cube trays work well for freezing eggs if they're used within a couple of

weeks. Some cooks prefer to freeze yolks separately from whites, but I prefer to stabilize eggs that have been lightly mixed with a fork or whisk. For every pint of raw egg (which amounts to about 10 whole eggs) I add about 1 to 1 1/2 teaspoons of sugar or 1 teaspoon of salt (depending on how they'll be used later). Frozen eggs work well for baking and cooking, especially quiches, but the whites will not produce meringue with stiff peaks.

LEFT: Organic eggs are ready to be stored at Fat Rooster Farm. The eggs are sold at the farmers' market and local co-ops.

Shallots hang to dry in a barn at Tunbridge Hill Farm. *"The Joy of Cooking* calls them the Queen of the Sauce Onions," farmer Wendy Palthey says.

State Resources

Alabama

Alabama Cooperative
Extension System
109-D Duncan Hall
Auburn University, AL 36849
334-844-4444
www.aces.edu

Tuskegee University

Cooperative Extension
1200 West Montgomery Rd
Tuskegee, AL 36088
334-727-8806
www.tuskegee.edu/
Global/category.
asp?C=34630&nav=CcXE

Alaska

Cooperative Extension Service
308 Tanana Loop, Room 101
University of Alaska Fairbanks
Fairbanks, AK 99775-6180
907-474-5211
cesweb@uaf.edu
www.uaf.edu/coop-ext

Arizona

University of Arizona
Forbes Building, Room 301
P.O. Box 210036
Tucson, AZ 85721
520-621-7205
Extension.arizona.edu

Arkansas

Cooperative Extension Service
2301 South University Avenue
Little Rock, Arkansas 72204
501-671-2000
www.uaex.edu

University of Arkansas at Pine Bluff

1890 Cooperative Extension
Service
1200 N. University Drive
Pine Bluff, AR 71601
870-575-8530
www.uapb.edu

California

El Dorado County
Cooperative Extension El
Dorado County
University of California at
Davis
311 Fair Lane
Placerville, CA 95667
530-621-5502
ceeldorado@ucdavis.edu
ceeldorado.ucdavis.edu

Colorado

Colorado State University
Extension
Campus Delivery 4040
Fort Collins, CO 80523-4040
970-491-6281
www.ext.colostate.edu

Connecticut

W.B. Young Building, Room 231
1376 Storrs Rd, Unit 4143
Storrs, CT 06269-4134
860-486-9228
extension@uconn.edu
www.cag.uconn.edu/ces/ces/
index.html

Delaware

University of Delaware
Ulysses S. Washington Center
1200 N. DuPont Highway
Mail Code D160
Dover , DE 19901-2227
302-857-6424
ag.udel.edu/extension

Florida

University of Florida
Department of Dairy and
Poultry
Sciences
P.O. Box 110920
Gainesville, FL 32610
352-392-1981
edis.ifas.ufl.edu

Georgia

Cooperative Extension
University of Georgia
111 Conner Hall
Athens, GA 30602-7505
706-542-3842
caesext@uga.edu
www.caes.uga.edu/extension

Hawaii

Cooperative Extension Office
University of Hawaii at Manoa
3050 Maile Way, Gilmore 203
Honolulu, HI 96822
808-956-8139
extension@ctahr.hawaii.edu
www.ctahr.hawaii.edu/site/
extprograms.aspx

Idaho

University of Idaho
Animal Veterinary Science
Department
Agricultural Science Building
Moscow, ID 83844-2330
208-885-6347
extension@uidaho.edu
www.extension.uidaho.edu

Illinois

Extension and Outreach
University of Illinois
214 Mumford Hall MC-710
1301 W. Gregory Drive
Urbana, IL 61801
217-333-5900
rhoeft@illinois.edu
www.extension.uiuc.edu

Indiana

Purdue Extension
Agricultural Administration
Building
615 W. State Street
West Lafayette, IN 47907-2054

765-494-8491
www.ag.purdue.edu/extension

Iowa

Iowa State University
Extension
Iowa State University
218 Beardshear Hall
Ames, IA 50011
515-294-4603
www.extension.iastate.edu

Kansas

Kansas State University
Research and Extension
123 Umberger Hall
Manhattan, KS 66506
785-532-5820
www.ksre.k-state.edu/
desktopdefault.aspx

Kentucky

University of Kentucky
College of Agriculture
S-107 Agricultural Science
Building
- North
Lexington, KY 40546-0091
859-257-4302
Darlene.mylin@uky.edu
ces.ca.uky.edu/ces

Louisiana

Louisiana State University
Agricultural Center
P.O. Box 25203

Baton Rouge, LA 70894
225-578-4161
www.lsuagcenter.com

Maine

University of Maine
Cooperative Extension
5741 Libby Hall
Orono, ME 04469
207-581-3188
Wwww-questions@umext.
maine.edu
www.extension.umaine.edu

Maryland

University of Maryland at
College
Park
Department of Animal and
Avian
Sciences
College Park, MD 20742
301-405-8746
extension.umd.edu

Massachusetts

University of Massachusetts
Extension Office
Draper Hall
University of Massachusetts
Amherst, MA 01003
413-545-4800
www.umassextension.org

Michigan

Michigan State University
Extension

Agriculture Hall, Room 108
East Lansing, MI 48824
517-355-2308
www.msue.msu.edu/portal

Minnesota

University of Minnesota
Extension
Office of the Director
240 Coffey Hall
1420 Eckles Avenue
St. Paul, MN 55108
612-624-1222
mnext@umn.edu
www.extension.umn.edu

Mississippi

Alcorn State University
1000 ASU Drive, #690
Lorman, MS 39096
601-877-6137
www.asuextension.com/asuep/
index.php

Mississippi State University

Department of Poultry
Science
P.O. Box 5188
Mississippi State, MS 39762
662-325-3416
www.msucares.com

Missouri

University of Missouri
309 University Hall
Columbia, MO 65211

573-882-7754
extension.missouri.edu

Montana

Montana State University
P.O. Box 172040
Bozeman, MT 59717-2040
406-994-1750
www.msuextension.org

Nebraska

University of Nebraska
UNL Extension
211 Agriculture Hall
Lincoln, NE 68583
402-472-2966
ltempel1@unl.edu
www.extension.unl.edu

Nevada

University of Nevada at Reno
Cooperative Extension
Reno MS 404
Reno, NV 89557
775-784-7070
www.unce.unr.edu

New Hampshire

University of New Hampshire
Cooperative Extension
Taylor Hall
59 College Road
Durham, NH 03824
603-862-1520
extension.unh.edu

New Jersey

Rutgers, The University of New
Jersey
Rutgers Cooperative Extension
88 Lippman Drive
New Brunswick, NJ 08901
732-932-5000
njaes.rutgers.edu/extension

New Mexico

New Mexico State University
Department 3AE
P.O. Box 30003
Las Cruces, NM 88003
505-646-3016
extension.nmsu.edu

New York

Cornell University
365 Roberts Hall
Ithaca, NY 14853
607-255-2116
www.cce.cornell.edu

North Carolina

North Carolina A&T State
University
P.O. Box 21928
Greensboro, NC 27420
336-334-7691
www.ag.ncat.edu/extension

North Carolina State University

Extension Poultry Science
229 Scott Hall

Campus Box 7608
Raleigh, NC 27695-7608
919-515-2621
www.ces.ncsu.edu

North Dakota

North Dakota State University
315 Morrill Hall
P.O. Box 6050
Fargo, ND 58105
701-231-8944
www.ag.ndsu.edu/extension

Ohio

Ohio State University
Extension Administration
2120 Fyffe Road
3 Agricultural Administration
Building
Columbus, OH 43210
614-292-6181
extension.osu.edu

Oklahoma

Oklahoma State University
Division of Agricultural
Sciences and
Natural Resources
136 Agricultural Building
Stillwater, OK 74078-6051
405-744-5398
www.oces.okstate.edu

Oregon

Oregon State University
101 Ballard Extension Hall
Corvallis, OR 97331-6702

541-737-5066
extension.oregonstate.edu

Pennsylvania

Pennsylvania State University
201 Agricultural
Administration Building
University Park, PA 16802
814-865-2541
extension.psu.edu

Rhode Island

University of Rhode Island
9 E. Alumni Avenue, Room 137
Kingston, RI 02881
401-874-2900
www.uri.edu/ce/index1.html

South Carolina

Clemson University
Clemson, SC 29634-0361
864-656-3311
www.clemson.edu/extension

South Dakota

South Dakota State University
Extension Director
P.O. Box 2207D
Brookings, SD 57007
605-688-4792
Latif.lighari@sdstate.edu
sdces.sdstate.edu/

Tennessee

University of Tennessee
Extension
2621 Morgan Circle

Starting with stone from a neighbor's building project, Laina Karoli and Claude Richter built a 7-foot by 8-foot root cellar into their hillside at True 4 Now Farm. The concrete roof was a slab they purchased; in all, the project cost $3,000 and Karoli hopes the space can be shared with other families in the community.

121 Morgan Hall
Knoxville, TN 37996
865-974-7114
www.utextension.utk.edu

Texas

Texas A&M University
Texas AgriLife Extension
Service
7101 Tamu
College Station,
TX 77843-2472
979-845-7800
agrilifeextension@ag.tamu.edu
texasextension.tamu.edu

Utah

Utah State University
4900 Old Main Hill Road
Logan, UT 84322
435-797-2200
extension.usu.edu

Vermont

University of Vermont
Burlington, VT 05405
802-656-3131
www.uvm.edu/extension

Virginia

Virginia Cooperative Extension

101 Hutcheson Hall (0402)
Virginia Polytechnic Institute
and
State University
Blacksburg, VA 24061
www.ext.vt.edu

Washington

Washington State University
421 Hubert Hall
P.O. Box 646230
Pullman, WA 99164
509-335-4561
ext.wsu.edu

West Virginia

West Virginia University
P.O. Box 6031
Morgantown, WV 26506-6031
304-293-5691
Extension.service@mail.wvu.
edu
www.wvu.edu/~exten

Wisconsin

University of Wisconsin
Department of Animal
Sciences
Animal Science Building
1675 Observatory Drive
Madison, WI 83706-1284
608-263-4300
servicecenter@uwex.uwc.edu
www.uwex.edu/ces

Wyoming

Cooperative Extension Service
Department 3354
University of Wyoming
1000 E. University Avenue
Laramie, WY 82071
307-766-5124
glen@uwyo.edu
ces.uwyo.edu

LEFT: **Donna Kausen pan-fries
shrimp for an appetizer on her wood
stove. The seafood was caught by
a local fisherman; extra fish can be
frozen to enjoy another day.**

Onions are gathered
for distribution to CSA
members at Sunrise Farm
in Hartford, Vermont.

References

Aidells, Bruce. *Bruce Aidell's Complete Book of Pork: A Guide to Buying, Storing and Cooking the World's Favorite Meat.* New York: Harper-Collins, 2004.

Bubel, Mike, and Nancy Bubel. *Root Cellaring: Natural Cold Storage of Fruits and Vegetables.* North Adams, MA: Storey, 1979.

Burch, Monte. *Building Small Barns, Sheds, & Shelters.* Pownal, VT: Storey, 1983.

Davidson, Alan. *The Penguin Companion to Food.* New York: Penguin, 2002.

Fallon, Sally. *Nourishing Traditions: The Cookbook that Challenges Politically Correct Nutrition and The Diet Dictocrats.* Washington, D.C: NewTrends, 1999.

The Gardeners and Farmers of Terre Vivante. *Preserving Food without Freezing or Canning.* White River Junction, VT: Chelsea Green, 1999.

Hobson, Phyllis. *Raising a Calf for Beef.* Pownal, VT: Storey, 1976.

Katz, Sandor Ellix. *Wild Fermentation: the Flavor, Nutrition, and Craft of Live-Culture Foods.* White River Junction, VT: Chelsea Green, 2003.

Maynard, Donald N., and George J. Hochman. *Knott's Handbook for Vegetable Growers*, fourth edition. New York: John Wiley and Sons, 1997.

McClure, Susan, and the Staff of the Rodale Food Center. *Preserving Summer's Bounty: a Quick and Easy Guide to Freezing, Canning, Preserving and Drying What You Grow.* Emmaus, PA: Rodale, 1998.

Mettler, John J., DVM. *Basic Butchering of Livestock and Game.* Pownal, VT: Storey, 1986.

Peery, Susan Mahnke, and Charles G. Reavis. *Home Sausage Making: How-to Techniques for Making and Enjoying 100 Sausages at Home.* North Adams, MA: Storey, 2003.

Ruhlman, Michael, and Brian Polcyn. *Charcuterie: The Craft of Salting, Smoking and Curing.* New York: W.W. Norton, 2005.

Schneider, Elizabeth. *Uncommon Fruits and Vegetables, a Commonsense Guide.* New York: William Morrow, 1986.

Shephard, Sue. *Pickled, Potted, and Canned: How the Art and Science of Food Preserving Changed the World.* New York: Simon & Schuster, 2006.

Tannahill, Reay. *Food In History.* New York: Stein and Day, 1973.

United States Department of Agriculture. *Complete Guide to Home Canning, Preserving and Freezing.* Mineola, NY: Dover, 1994.

Whealy, Kent, *Garden Seed Inventory,* Decorah, Iowa: Seed Savers Exchange, 2005.

Buttercup squash are wrapped in
burlap and stored in the root cellar
at Fat Rooster Farm. The burlap
keeps light off of the squash, which
prolongs the length of storage.

Glossary

Acid Foods – Those foods that contain enough acid to keep a pH level of lower than 4.6 and can therefore be safely processed using a boiling water bath without the fear of botulism.

Altitude – How high a location is vertically with reference to sea level. Altitude affects the time required for proper canning.

Appertization – Several decades before Pasteur would discover bacteria, Nicolas Appert developed a way to bottle and preserve vegetables and fruits by sterilizing them in an autoclave. The method used extremely high temperatures to seal the bottles; the Frenchman became known as the Father of Canning.

Ascorbic acid – The proper chemical name for vitamin C, used to prevent discoloration of some fruits and vegetables after they've been peeled. Lemon juice contains high amounts of ascorbic acid and is commonly used.

Bacteria – A group of one-celled microorganisms, first discovered by Louis Pasteur that can grow and divide rapidly in suitable environments. Undesirable bacteria can cause food spoilage and disease, while some are used to manufacture antibiotics or intentionally ferment foods to make finished products, like wine, beer or yogurt.

Bail-top – An old-fashioned style canning jar which had a wire bail that would secure the glass lid and rubber ring close to the mouth of the jar. The jars are not considered safe for canning uses in the United States, although they're still widely used in other parts of the world.

Blanching – A technique that is sometimes used to prepare foods prior to being canned or frozen. The process helps loosens skins on fruits and vegetables and it also helps to cease microbial and enzymatic activity, and to retain color and texture of the food.

Boiling-water bath – Also referred to as a **boiling water canner** or **hot water bath**, this method of canning uses boiling water to seal canning jars that are placed inside it. It is only suitable as a canning method for foods that are high acid foods, like pickles, jellies, and most fruits and tomatoes.

Brining – A method of food preparation that uses salt, water and on occasion sugar and a variety of herbs and spices. Poultry

products and pork products are most often brined; it renders the meat more tender, flavorful and often more juicy. Brines are also used to prepare cuts of meats intended for later smoking.

Botulism – An illness caused by consuming a toxin produced by the bacteria *Clostridium botulinum* that is potentially fatal at miniscule amounts. Proper growth conditions for the bacteria are moist, low-acid foods that have been preserved in an environment with little or no oxygen. Proper processing with the correct amount of heat, high acid environments, freezing, drying, or high amounts of oxygen in fresh foods all prevent the disease.

Canning Jar – A glass container specifically manufactured for preserving foods through canning in a boiling-water bath or a pressure canner. Jars range wildly in size and shape. Jars that previously contained foods purchased at the supermarket should not be reused for canning; the glass is thinner than jars manufactured for home canning and may shatter upon reuse.

Canning Ring – A term referring to the lid fits around the canning jar lid and secures the lid to the jar. After processing and sealing, the rings can be safely removed and reused, and the lid will remain in place and continue to keep the food sealed while on the shelf. If the jars are moved frequently, rings should stay on the lids to prevent accidental vacuum loss of the jar and subsequent food spoilage.

Canning Salt – Salt that has no anticaking or iodine added; kosher, sea salt and pickling salt are all synonymous.

Citric acid – A type of acid commonly added to canned foods to increase the acidity and reduce the chance of spoilage and food borne illness. Citric acid also enhances flavor and improves the color of some canned foods.

LEFT: Delicata squash cure in a stall in the barn at Fat Rooster Farm. With its edible skin, Delicata doesn't store as well as other varieties of winter squash.

Clostridium – One genus of bacteria known to cause food borne illnesses. *Clostridium botulinum* can produce a toxin which causes the disease called Botulism, given the right conditions.

Cure – A process used to prepare foods that are intended for smoking or drying and usually consist of salt, sugar and herbs and spices. Cures can be wet, most commonly referred to as brines, or dry-cures (also referred to as **dry rubs**).

Drying – A very simple and ancient form of food preservation wildly used worldwide for foods of all kinds. By reducing the water content within the food, micro-organisms and enzymes that cause food spoilage are prevented from existing. Drying can also be combined with salting and smoking.

Enzymes – Proteins that are present in foods that speed up changes in the color, texture, flavor and nutritional value of the food, particularly after it has been cooked, cut, sliced or crushed. Blanching foods prior to freezing or canning will suspend enzymatic activity and improve the quality of the preserved product.

F1 hybrid – A variety of fruit or vegetable that is the result of crossing two distinctly different varieties or parents to create the first generation of seeds, or filial 1 generation. The offspring that are produced have combined traits from the parent stock and will not produce seed that is true to either parent or to itself. The opposite of F1 hybrids are open-pollinated varieties that can be saved and used to produce seeds and plants that resemble the seedstock.

Glass lid – An old fashioned canning lid that is held in place with a rubber ring and a bail to seal the jar. These jars are considered unsafe for canning in the United States, but are still widely used in other parts of the world.

Headspace – The space in the canning jar between the food or liquid and the top of the jar that is needed for food expansion after the food is heated so that a vacuum can form and properly seal the jar.

Hot pack – A canning term that refers to pre-heating food with steam or boiling water before it is put in jars and sealed.

Jar up – A colloquial, old-fashioned term for canning, meaning to put food into and seal by canning in a boiling-water bath.

Whitney Taylor of Wellsboro, Pennsylvania, left, and Tali Biale of New York City pull squash vines from the garden before garlic is to be planted at Fat Rooster Farm. Garlic plants are rotated every season on the farm.

Lacto-fermentation — An ancient food preservation that is the precursor to pickling. Lactic acid-forming bacteria create conditions inhospitable to pathogenic bacteria and food spoilage by creating a high acid environment through fermentation. The most common lactic acid-forming bacteria are the lactobacilli. The bacteria not only preserve the food, but they enhance it nutritionally by working with enzymes to make the food more easily digestible and higher in vitamins. Sauerkraut is the most popular form of lactofermented vegetables, though the art is regaining in popularity.

Low acid foods — Any foods that have a pH higher than 4.6, meaning they contain very little acid and are much more likely to harbor the potential for growth of *Clostridium botulinum* which causes the toxin responsible for Botulism. Meets, fish, figs, sea-foods, some dairy and some tomatoes are all low-acid foods and require processing in a pressure canner or adification before preserving safely can be achieved.

Listeria – A bacterium that can cause the food borne illness listeriosis. Pregnant woman and others with pre-existing illnesses and impaired immune systems are most susceptible to the disease. The bacteria are killed with high heat, so heating foods prior to eating will render them unharmful. Prepared salads, luncheon meats and soft cheeses are often associated with listeriosis.

Mold – A microorganism that is actually a type of fungus and occurs wildly throughout the natural world. Some molds are encouraged for growth to preserve foods, like cheeses. Others are used to manufacture antibiotics. They are usually visible and often colorful. Correct sealing during canning processes can prevent the growth of molds on preserved foods.

Mycotoxin – A toxic by-product produced by some forms of mold on foods.

Open kettle canning – A form of canning no longer recommended due to the risk of spoilage and food borne illnesses. Food was typically heated in a covered kettle then poured hot into jars and sealed. Jams and jellies were often canned using this method. The risk of recontamination of the food when it is poured into the jars as well as the low vacuum created by simply pouring rather than processing sealed jars does not yield consistent, safe food.

Open-pollinated – Refers to plants that produce seeds when they've been pollinated by wind, insects, birds or other natural mechanisms and that have the potential to recreate plants with similar traits as the parents.

Pasteurization – The act of heating food to a temperature that will kill heat-resistant microorganisms known to be found in that food.

Pickling – A technique that uses vinegar and salt to create a high acid environment to safely preserve vegetables and meats. Pickled products are often processed and sealed or kept refrigerated until consumed.

Pickling spices – A mixture of herbs and spices, often sold already prepared. Black pepper, celery seed, coriander and dill are some of the more common ingredients.

pH – A scientific measure of acidity and/or alkalinity. A food with a neutral pH is 7.0; if lower, the food is considered acidic; higher values are considered alkaline. The values range from 0 to 14; low-acid foods are all those with values above 4.6 and therefore run the risk of causing food borne illnesses if improperly preserved.

Pressure canner – A metal pot designed for processing low-acid foods at high heat using increased pressure. The lids are lockable and designed to regulate the internal pressure with an exhaust valve. A **pressure cooker** is smaller than a pressure canner and does not fit canning jars conveniently.

Raw pack – To fill jars with raw, unheated food, typically used when canning low-acid foods. High acid foods that are processed in a boiling-water bath tend to have less loss of quality and nutritional value.

Rubber rings – The seal between the canning jar and the glass lid found on old-fashioned canning jars (still available for sale in many areas of the United States and throughout the world).

Salmonella – A group of bacteria that are responsible for one of the commonest causes of food borne illness, often associated with spoiled or tainted poultry products, but also found in other foods. While unpleasant, the disease is not usually fatal to healthy humans. Proper cooking of foods will kill the bacteria.

Salting – An ancient form of food preservation that uses salt, salt petre, nitrates or nitrites to deplete excess liquid from the food's cells, creating an uninhabitable environment for food spoiling bacteria.

Smoke – An important adjunct to salting and drying, smoking foods also adds flavor. In antiquity, the tannins that clung to the smoked foods would protect the outer layer from insects because of their bitterness and impenetrable film on the meat.

Yeasts – Microorganisms that are used to ferment and preserve some foods.

Due to a wet summer, the author had enough artichokes to can—shown are pickled globe artichokes.

Index

Italics indicate photo captions.